CHEMICAL KIM
KALAMAZOO VALLEY COMMUNITY COLLEGE

CHEMISTRY AT THE BEGINNING

Kendall Hunt
publishing company

Cover image: © Shutterstock, Inc.
Inside cover image provided by the author.

www.kendallhunt.com
Send all inquiries to:
4050 Westmark Drive
Dubuque, IA 52004-1840

ISBN 978-1-4652-7573-8

Printed in the United States of America

CONTENTS

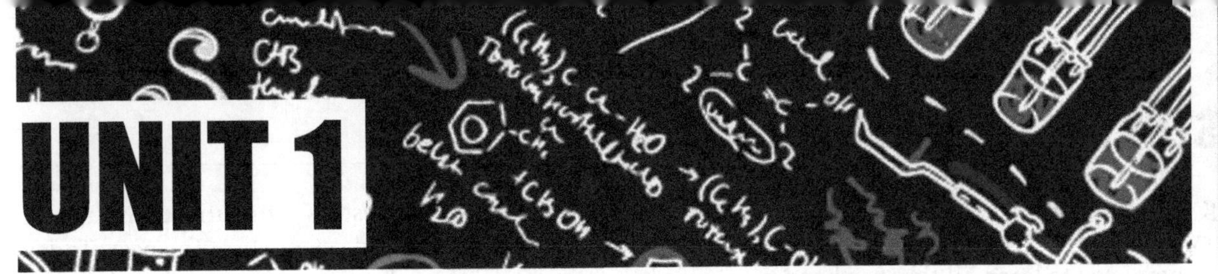

ALL ABOUT MATTER (MATTER AND ENERGY)

Our Study of Chemistry Begins

Possible Careers in Chemistry

Definition of Chemistry

History of Chemistry

- Chemistry Timeline
- Aristotle
- The Philosopher's Stone
- Chemistry Organized
- Describing Material
- Phlogiston
- The Discovery of Phosphorus

Introduction—Matter

States of Matter

Physical and Chemical Properties

Physical and Chemical Changes

Classifying Material (Matter)

Chemical Symbol

Chemical Equations

Energy in Chemical Change

All About Matter

"The chemists are a strange class of mortals who seek their pleasure among soot and flame, poisons and poverty, yet among all these evils I seem to live so sweetly that may I die if I would change places with the Persian king."
~John Joachim Becher

OBJECTIVES FOR UNIT 1

- To understand the science of chemistry.
- To understand the beginnings of chemistry.
- To understand the history of elements including Aristotle's four-element theory and the phlogiston theory.
- To understand and apply the terms: macroscopic, microscopic, particulate, and states of matter.
- To understand the differences between physical and chemical properties and physical and chemical changes.
- To be able to classify material as one or more of the following: heterogeneous, homogeneous, solution, pure substance, compound, and/or element.
- To understand the formula for a chemical compound and the equation for a chemical reaction.
- To understand energy: endothermic vs. exothermic and potential vs. kinetic.
- To learn some basic conservation and energy laws.

OUR STUDY OF CHEMISTRY BEGINS

Entering into the study of chemistry is like walking on the shoulders of ancestors. Our understanding of chemistry evolved from philosophers, alchemists, inventors, nurses, doctors, and scientists of the past. Today we study chemistry to learn all that has been discovered about matter. We seek answers to everyday phenomena and create questions to new observations. Chemistry is a fascinating science that involves studying life, environment, medicines, atmosphere and space, synthetics, and many other wonderful disciplines.

POSSIBLE CAREERS IN CHEMISTRY

Careers include: agricultural research, anti-cancer research, biochemistry, biotechnology, biomaterials, brewing industry, building research, chemical analysis, chemical technology, dairy industry, dentistry, pharmacy, electric power generation, engineering, environmental chemistry, fertilizer industry, food and drink technology, forensic science, forestry research, health and safety, horticultural research, industrial chemistry, investment banking, laboratory manager, light-emitting polymer research, local council work, marine sciences, materials science, medical laboratory testing, medicine, microbiology, pesticides industry, petrochemical industry, pharmaceutical industry, plastic products science, polymer industry, pyrotechnics, quality control, regional council, research, science publishing, school teaching, steel industry, water treatment plant, water-quality analysis, and wine industry.

DEFINITION OF CHEMISTRY

Chemistry is the study of matter (material) and energy. Matter is defined as anything that has mass and occupies space. If you break matter down, it will be comprised of one or more of the chemical elements on the periodic table. Energy is the ability this material has to do work.

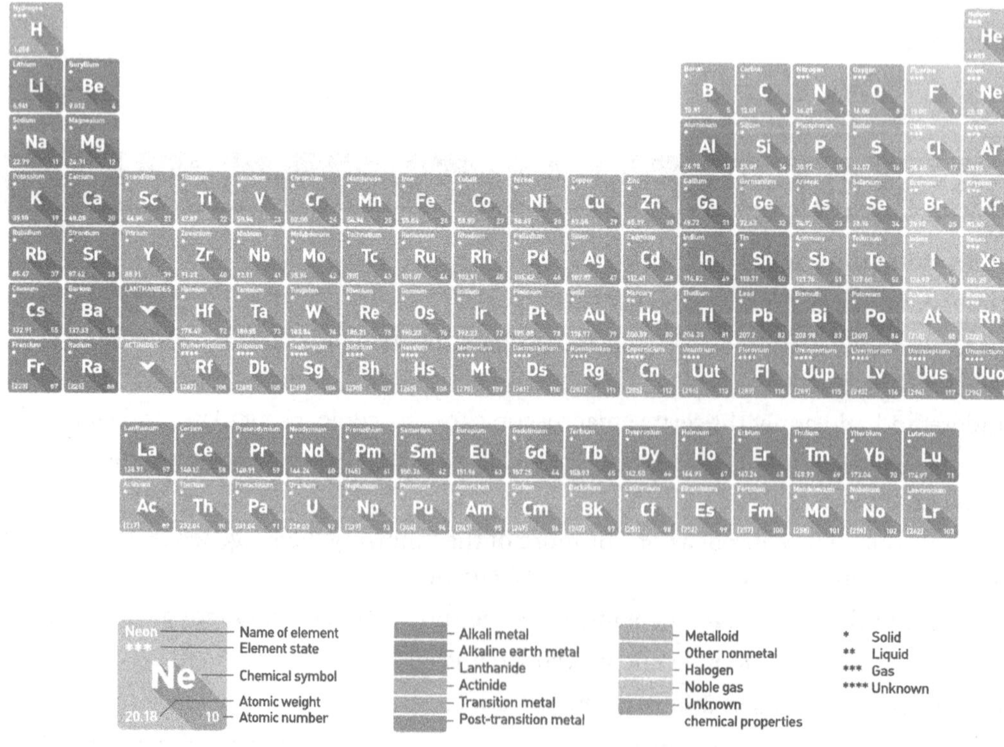

HISTORY OF CHEMISTRY

Chemistry has evolved as humans have evolved. We have learned about matter and its behaviors from our needs to eat, hunt, find shelter, farm, cook, fight, and stay healthy. Much about chemistry that is studied in American schools comes from European nations. However, it is important to note that many other nations have been contributors to our modern understanding of matter and energy. European authors used works from other discoverers around the world along with their own knowledge to advance chemistry.

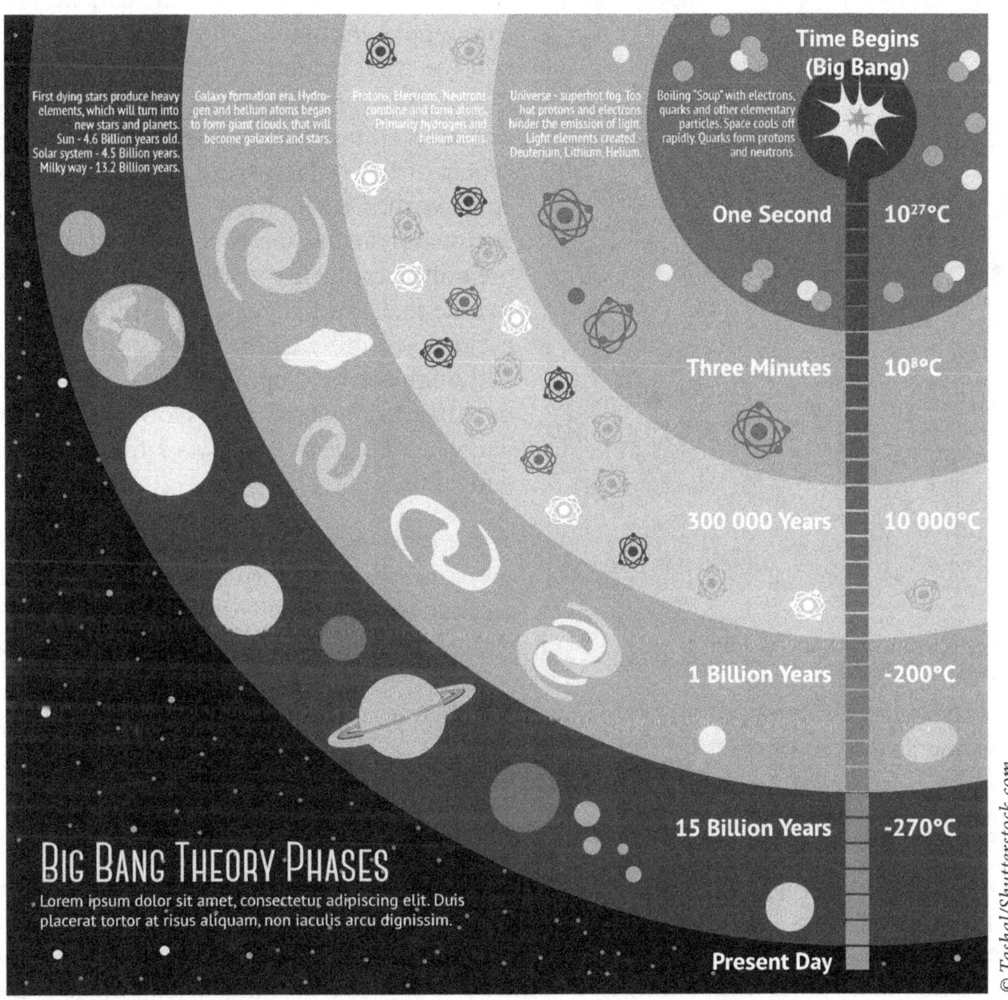

CHEMISTRY TIMELINE

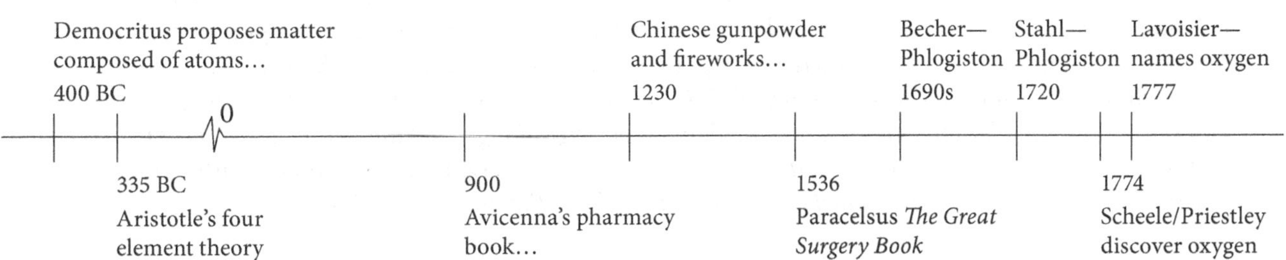

ARISTOTLE

Early Greek philosophers who relied on observations instead of experimentation in explaining the behaviors of basic materials were the first to write about chemistry. Around 400 BC, one such philosopher named Democritus stated that, when broken down, all material is comprised of atoms. Shortly after 335 BC, another better-known philosopher named Aristotle wrote a different theory about material in which matter is composed, not of atoms, but of the four elements of water, earth, air, and fire. This became known as the four element theory of matter and was held to be the only acceptable theory about the composition of matter for almost 2,000 years in most of Europe.

© tulpahn/Shutterstock.com

Aristotle also proposed a fifth element, *aether* (for the sun, moon, and stars), from which came the word *ethereal* (celestial or airy). In the Middle Ages, Aristotle's theory became the official philosophy of the Roman Catholic Church. Aristotle's theory made earth the center of the universe and the heavens obeyed different laws than earth. Most of Europe did not advance in chemistry during the time that Aristotle's theory was the only acceptable theory about matter. It is important to note that advancements in chemistry did occur other nations including China, Egypt, and Islamic countries. Although the developments were brought into Europe, they had to remain secret. However, they gave us such names as *khemeia* (root word for chemistry), alcohol, and elixir.

Those who initially practiced khemeia were those who prepared the dead for the afterlife. Khemeia was later associated with other processes connected with seven metallic elements: gold, silver, copper, tin, lead, and mercury. To keep secret the trade of khemeia, the seven planet names were used to disguise these elements. For example, For example, Venus and Jupiter (the secret names for copper and tin) were used to create bronze. Not surprisingly, those who practiced khemeia were called wizards, magicians, or sorcerers.

THE PHILOSOPHER'S STONE

For almost two centuries European's under the belief of Aristotle science performed experiments on matter in attempts to finding what is famously known as "seeking the philosopher's stone." These are experiments involving the technique of turning metal into gold (referred to as transmutation) as well as the ability to achieve immortality. Everything from taking lead metal and burying it in a hillside, hoping it would grow into gold, to performing numerous laboratory experiments of mixing, heating, pounding, dissolving, and tasting were tried.

> *"In the dark interior of an old laboratory cluttered with furnaces, crucibles, alembics, stills, and bellows, bends an old man in the act of hardening two thousand hens' eggs in huge pots of boiling water. Carefully he removes the shells and gathers them into a great heap. These he heats in a gentle flame until they are white as snow, while his co-laborer separates the whites from the yolks and putrefies them all in the manure of white horses. For eight long years the strange products are distilled and redistilled for the extraction of a mysterious white liquid and a red oil. With these potent universal solvents the two alchemists hope to fashion the 'philosopher's stone.'"* [1]

© Everett Historical/Shutterstock.com

In the 14th Century, a Frenchman named Nicolas Flamel gained fame by his attempts to find the philosopher's stone. Although it is recorded that Flamel died at the age of 88, a legend grew up that he achieved eternal life as a result of his experiments. His fame continues to this day and he is famously depicted in the Harry Potter book series as the wizard who created the philosopher's stone (called the sorcerer's stone in the American versions).

People who performed experiments seeking the philosopher's stone were referred to as alchemists. Alchemistry is a word derived from Egyptian and Arabic languages and dating back more than 1500 years. With our understanding of chemistry today, one might think that efforts to seek the philosopher's stone was a waste of time. Were these past seekers of wealth and eternal life contributing nothing to our modern sciences? Were their only contributions to our modern society additions to fictional stories such as Harry Potter? Maybe these people were not scientists, but were witches and wizards? What did alchemists achieve for our modern sciences? It will be through our investigations of chemistry that we will find the answer.

One scientist in 1980 had such a good understanding of chemistry, he did succeed in turning lead to gold. Following is an article from Guinness World Records—*Lead Turned into Gold*:

> In 1980, the renowned American scientist Glenn Seaborg (1912–1999) transmuted several thousand atoms of lead into gold at the Lawrence Berkeley Laboratory. His experimental technique, using nuclear physics, was able to remove protons and neutrons from the lead atoms. Seaborg's technique would have been far too expensive to enable routine manufacturing of gold from lead, but his work is the closest to the mythical Philosopher's Stone.[1]

© Nerthuz/Shutterstock.com

CHEMISTRY ORGANIZED

The Persian (modern-day Iran) philosopher, Abu Ali ibn Sina or Avicenna (980–1037) was unique in his approach to bringing order to science. Avicenna classified everything from medicines to experiments which he recorded in a book titled, *al-Qanun*. This book was an extensive encyclopedia for its time. It was considered the first pharmacy book and remained an important source throughout Western civilization for six centuries. Just as Avicenna discovered, classifying brings more order to the understanding of concepts. Today, we continue this process of classification of matter.

© YANGCHAO/Shutterstock.com

DESCRIBING MATERIAL

A famous alchemist by the name of Paracelsus lived in the 16th century. His contributions to chemistry are numerous. He was the first to realize that zinc was metallic and to describe the properties of bismuth and cobalt (although he didn't realize these were elements). Around this time antimony was discovered. Middle Eastern women darken their eyebrows with antimony to increase their seductiveness. It may be that the word antimony is derived from the word *anti-monakhos* or anti-monks because a German monk named Basil Valentinus (yes, perhaps Valentine) fed antimony to his fellow monks in the hopes of getting them to gain weight. Unfortunately, it led to their deaths. Jezebel in the bible is mentioned as using a similar substance referred to as *kohl* (an Arabic word). The word kohl was to describe distilled liquids, thus *al-kohl* (or alcohol).

PHLOGISTON

In the late 1600s, a German chemist named Johann Becher presented a theory to explain the mystery of fire. Becher asked such questions as what happens to paper when it burns and what happens to metal when it rusts? Becher proposed when substances burn, they contain an earth element called **phlogiston** (from the Greek word *phlogios*, meaning fiery). The phlogiston theory started to move chemistry into a new direction. Unknowingly, Becher's phlogiston theory initiated the first major challenge to Aristotle's four element theory.

At the beginning of the 1700s, another German professor of medicine named Georg Stahl did further work on the phlogiston theory. The Becher/Stahl theory explained burning, oxidation, calcinations (metal residue after combustion), and breathing in the following way[2]:

- Flames extinguish because air becomes saturated with phlogiston.
- Charcoal leaves little residue upon burning because it is nearly pure phlogiston.
- Mice die in airtight space because air saturates with phlogiston.
- When heated, metals are restored because phlogiston transferred from charcoal to calx.
- Air is the carrier of phlogiston.

This phlogiston theory leads to one of the most important discoveries of the time—oxygen.

Take yourself back to the turn of the century (the 18th century that is). Walk around in one of the chemistry laboratories set up throughout the Western world and you will observe the many different gases being released, breathed, and coughed out. Having just learned about this phlogiston theory, you notice chemists analyzing these gases. You meet a Swedish chemist named Karl Scheele who tells you all about his tests and observations on gases. He tells you about one gas (you know as HCN or hydrogen cyanide) and how he tested it. You are amazed because you know that HCN can cause a painful death if inhaled or absorbed through the skin. Scheele also shares with you an amazing discovery, stating that air contains two gases he calls "fine air" (you know it as oxygen) and "spoiled air" (you know it as nitrogen). He shows you how he isolated the "spoiled air." Scheele uses the phlogiston theory to explain his findings.

At this same time, an Englishman named Joseph Priestly made the same discovery about air. Both Scheele and Priestly led us to the discovery of oxygen. But Aristotle's theory and the phlogiston's theory were holding strong, so neither Scheele nor Priestly announced the discovery of a new element. Priestly tested the newly discovered oxygen by breathing a sample, which he called "dephlogisticated air." He found that he enjoyed the "dephlogisticated air" very much.

© Georgios Kollidas/Shutterstock.com

Around 1774, Priestly passed his discovery along to a Frenchman named Antoine Lavoisier and his wife Marie-Anne. The Lavoisiers disproved of the phlogiston theory and knew immediately that Priestly had presented a new element. Antoine Lavoisier named this new element oxygen (Greek *oxy-* meaning acid and *-gen* meaning generator). Madame Lavoisier threw a big party to celebrate her husband's new discovery at which she ceremonially burned the works of Becher and Stahl, declaring the phlogiston theory was now dead in France. Many denote Antoine Lavoisier as the "Father of Chemistry" because of his forward thinking of chemical elements and compounds. One great historian wrote: "While out of its ashes, like the Phoenix of old, sprang up a new chemistry." [3]

THE DESCOVERY OF PHOSPHORUS

German alchemist Hennig Brand discovered phosphorous in 1669. His marriage to a wealthy woman allowed him to pursue investigations in alchemy. One such investigation involved the collection of fifty buckets of urine that he left out to evaporate and putrefy, eventually breeding worms. The urine continued to ferment until turning black. Brand heated this black substance, which produced a transparent waxy substance he collected under water. When he removed this substance from water, it glowed in the dark and sometimes even spontaneously ignited, giving off dense white fumes. He named this new substance phosphorus from the Greek *phos* (light) and *phoros* (bringing). Interestingly, phosphorus is the only element whose symbol describes how it was discovered.

What is the chemical symbol for phosphorus?

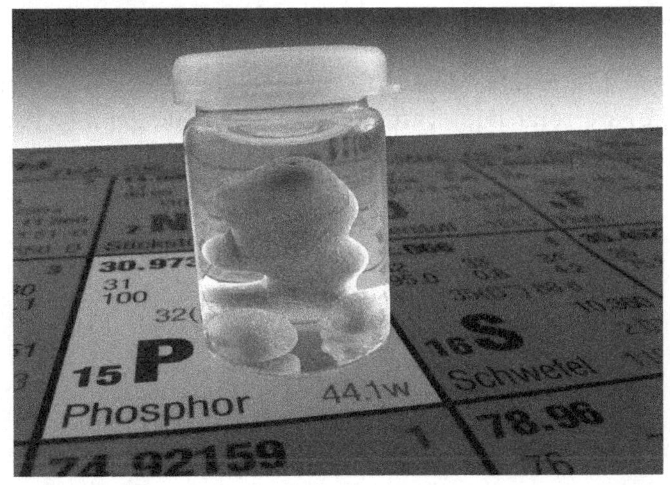

© magnetix/Shutterstock.com

INTRODUCTION—MATTER

Have you ever wondered?

"How is paper made?"
"How does my cell phone send pictures?"
"Why does my face always feel so oily?"
"How does aspirin find its way to my headache?"
"What exactly is fire made of?"

Beyond your wonderment, have you explored these thoughts? Have you examined a piece of paper with a magnifying glass? Looked inside your cell phone for the little Kodak™ man? Spent time staring into a fireplace?

What did you find? Did your examination answer your question?

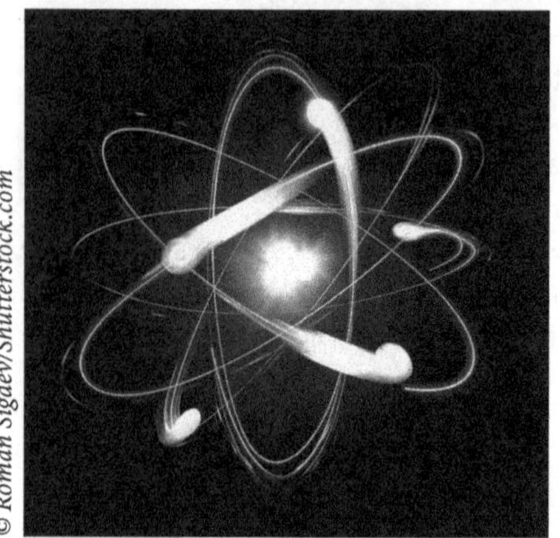

© Roman Sigaev/Shutterstock.com

Could you go a little further? Are there any tests you could perform?

What about things that motivate you? Do you have a desire to help others? Do you have a curiosity about nature? Are you determined to make a lot of money? If you answered yes to any one of these questions, you are a perfect candidate to study and understand chemistry.

Yes, chemistry, the science that reaches all sciences.

What is chemistry? Quite simply, chemistry is the study of matter and the changes that take place within that matter.

What is matter? **Matter** is the name scientists have given to everything that you can touch, or see, or feel, or smell. Anything that has mass and takes up space is matter! Everything on earth, everything in our solar system, everything in our galaxy, and everything in the universe is made up of matter.

1. **Macroscopic** matter is matter that can be seen with the human eye.
2. **Microscopic** matter is matter too small to be seen by the human eye, but can be seen with an optical microscope.
3. **Particulate matter** is matter too small to be seen by the human eye and too small to be seen with an optical microscope.

STATES OF MATTER

Well, let's see, there is Michigan, Minnesota, Montana...oh wait, not states of America but *states of matter.* We will study three (not fifty). They are: solid, liquid, and gas. The state of matter for a particular material depends upon the surrounding temperature, atmospheric pressure, and the specific characteristics of the particular type of material.

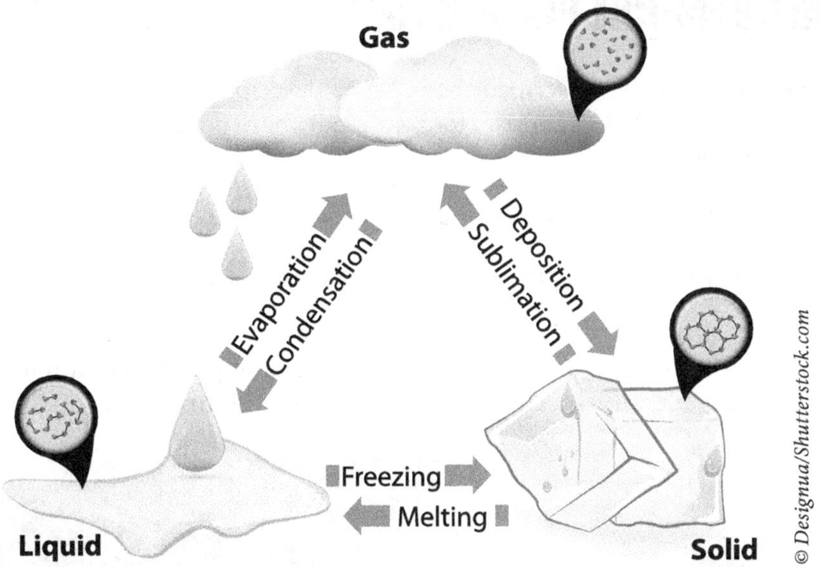

Definitions and Examples:

- **Solids** hold their shape. Examples include wood, steel, paper, copper.

  At the microscopic level, solids atoms have very little movement.
- **Liquids** will fill the shape of a container. A container is needed in order to "hold" a liquid. Examples include water, mercury, milk, pancake syrup.

  At the microscopic level, liquid atoms have more movement than solids, but they are still close together.
- **Gases** can fill a container of any shape or size because they do not hold their own shape. Some people may think that certain gases don't exist because they can't be seen, but they are material! Examples include oxygen, nitrogen, helium carbon dioxide, carbon monoxide.

  At the microscopic level, gas atoms have high movement and are not close together; they are only in collisions with one another.

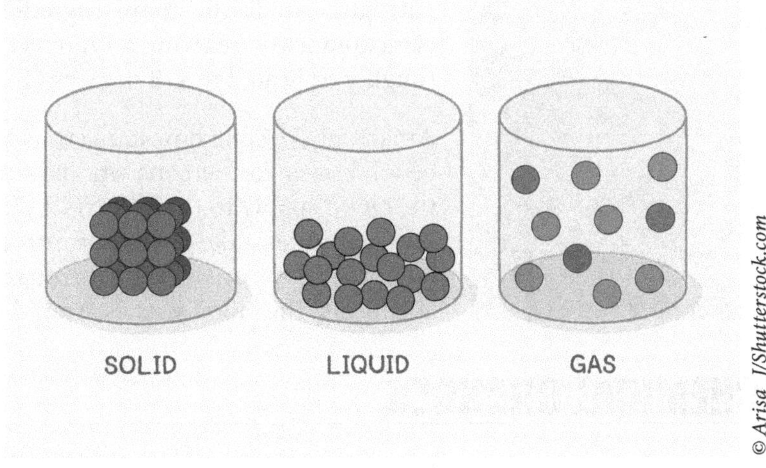

PHYSICAL AND CHEMICAL PROPERTIES

Upon discovery of new matter, people described them. Some descriptions included metallic, color, poisons, and spirits (distilled alcohol drinks). As chemists today, we put these description words into two categories: physical or chemical properties.

Physical properties are properties of a substance that depends upon the substance itself.

If the amount of silver used in making a silver cup were increased, the volume size of the cup would also increase.

Examples include mass, length, volume, physical state (solid, liquid, or gas at a certain temperature and pressure), color, texture, taste, odor, density, specific heat, melting point, boiling point, ductile, and conductivity.

© joreks/Shutterstock.com

© choikh/Shutterstock.com

The first person to scientifically explain a rainbow was a 13th Century monk by the name of Dietrich von Freiberg. He mimicked a large water drop by filling a spherical glass flask with water. When light passed through the flask, he observed its diffractive and reflective properties. This experiment produced what we know today as the rainbow effect.

Chemical properties are properties of a substance that depend upon the action of the substance in the presence of other substance.

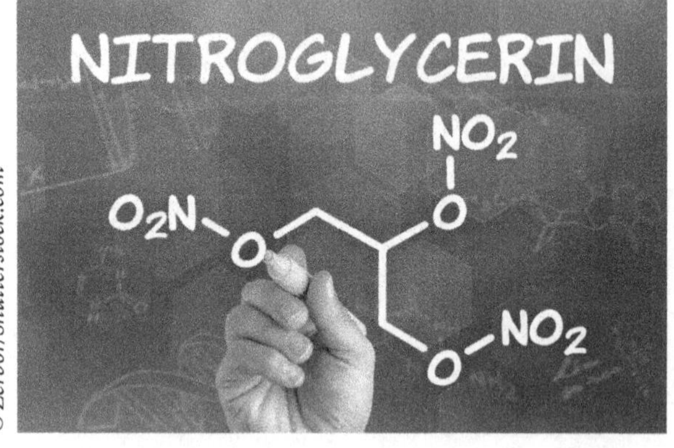

© Zerbor/Shutterstock.com

Examples:

Nitroglycerin is the main ingredient in dynamite and can explode (by reacting with oxygen) if it is handled too roughly or heated to 218 degrees C.

Amazingly, despite how dangerous nitroglycerin is, it can be a lifesaver for persons who have problems keeping up the blood supply to their hearts. When chest pain strikes, they will often take a dose of nitroglycerin, which helps by opening up or dilating (through reacting with body chemicals) the blood vessels around the heart.

PHYSICAL AND CHEMICAL CHANGES

How do you categorize the changes matter undergoes?

Let's take steel for example. One day I decided to investigate how the steel on my truck was made, so I visited a steel factory. At the factory I watched as iron was melted (turned into a liquid), purified, and mixed with other elements such as carbon and manganese, to form the material called steel. The final product was a shiny metallic substance. Hopping back into my truck, I noticed the steel on my truck was not the same shiny metallic substance made at the steel factory. The steel on my truck had turned a brownish color and was starting to become thinner in some spots. This is called rusting.

© jordache/Shutterstock.com

Melting and rusting are two of the changes that material undergoes. These changes can be described as physical and chemical changes.

Physical changes are changes in a substance that affects only the "appearance" of the substance. Examples include dissolving, filtering, evaporating, vaporizing, condensing, melting, boiling, and freezing.

Filtering a substance is a physical change in which two substances are separated without reacting and changing their composition chemically.

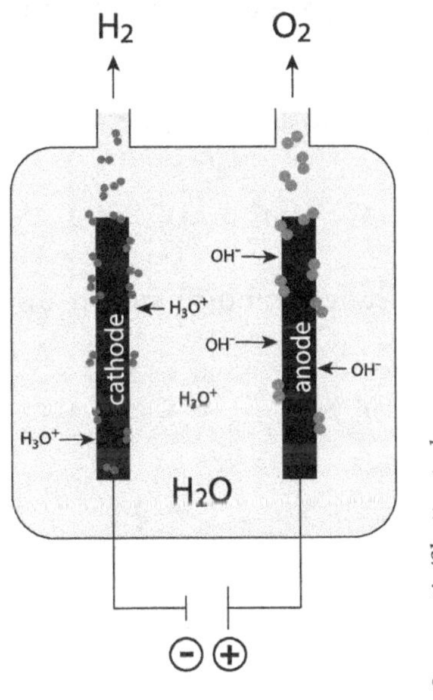

© magnetix/Shutterstock.com

© photosync/Shutterstock.com

Chemical changes are changes involving one or more new substances that are formed with different properties than the substance that undergoes the change. Examples include burning, tarnishing, corroding, exploding, digesting, oxidizing, and reacting.

When water and air (oxygen) come into contact with iron, iron rusts. The electrons in iron are moving from the metal to the water and oxygen, producing iron oxide. To prevent rust from occurring in water pipes, it is possible to attach a piece of metal magnesium to the pipes. The water and air then react with the magnesium and leave the iron pipe alone.

CLASSIFYING MATERIAL (MATTER)

1. Heterogeneous
2. Homogenous / Solution
3. Homogenous / Pure Substance / Compound
4. Homogenous / Pure Substance / Element

1. **Heterogeneous**

Heterogeneous materials are composed of two or more materials that have physically combined but are not uniformly mixed. A phase is any region with a uniform set of properties. The different phases in a heterogeneous material are separated from each other by definite boundaries called interfaces. Heterogeneous material can be physically separated. Heterogeneous materials are called mixtures and are not uniform throughout.

Examples:

Let's make a bottle of salad dressing using olive oil and spices. Shaking the bottle "mixes" the olive oil and spices. Allow the bottle to sit for a short period of time and a separation will take place. One could further use filtration to physically separate the olive oil from the spices.

Samples taken from two "different" phases in a heterogeneous mixture will have different properties.

2. **Homogenous / Solution**

Homogeneous / Solutions are composed of two or more materials that have physically combined and are uniformly mixed. A solution may be a solid, liquid, or gas.

Examples:

A solid: brass, a solution of tin and copper.

A liquid: seawater contains salt particles as the solute and water as the solvent.

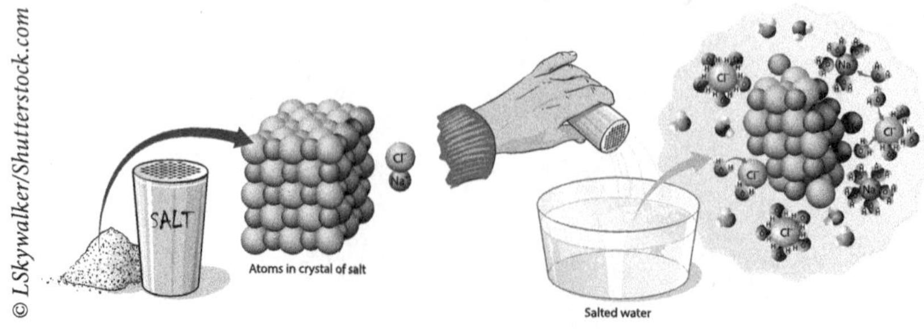

A gas: air, composed primarily of nitrogen and oxygen.

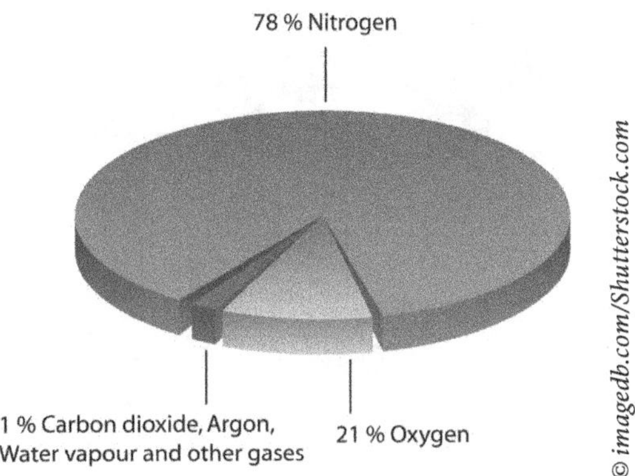

3. **Homogenous / Pure Substance / Compound**

As discussed earlier, Antoine Lavoisier did not agree with the phlogiston theory. He disproved this theory by proving the existence of compounds. Homogeneous / Pure Substance / Compounds are composed of more than one kind of atom and are written in a chemical formula. The first compounds that Lavoisier investigated were those that involved oxygen in a chemical change called combustion. Lavoisier heated lead metal in a closed container. He weighed the entire apparatus (containing both lead and air) both before and after the experiment and found no overall loss of weight. However, Lavoisier did find that the lead gained weight. Where did this extra weight come from? The air lost weight; therefore there was no phlogiston. What Lavoisier discovered was a compound containing lead and oxygen (oxygen from the air).

Chemical formulas are written to represent compounds.

Examples of compounds:

Water: H_2O

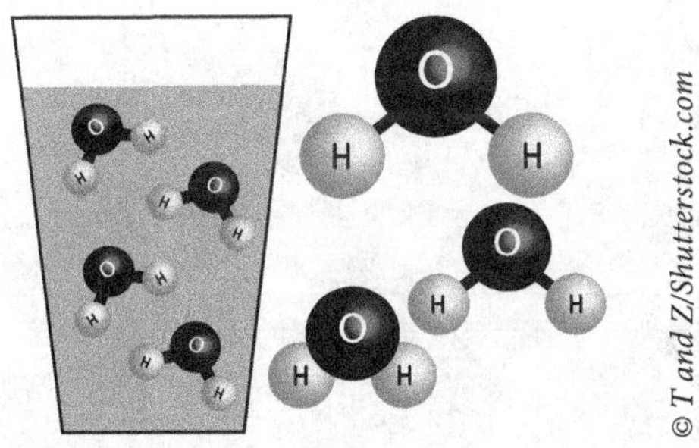

Table Salt: NaCl

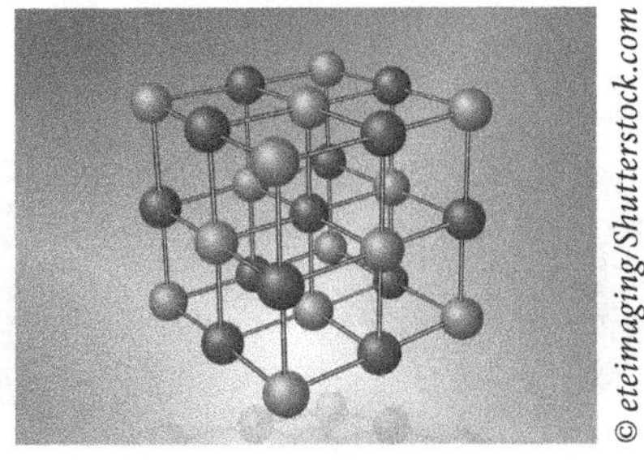

Glucose (simple sugar): $C_6H_{12}O_6$

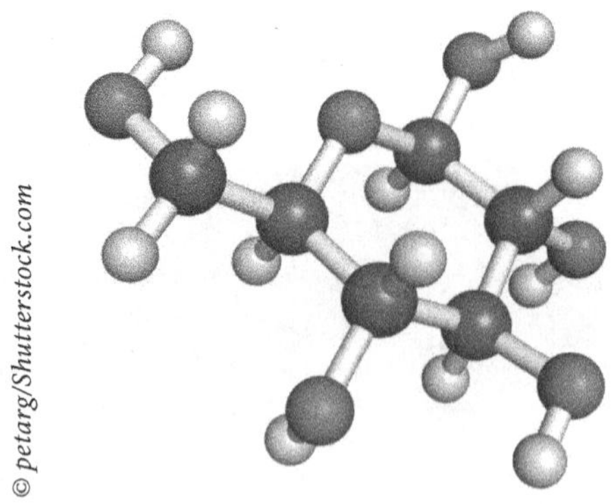

NH_3: 1 atom N, 3 atoms H

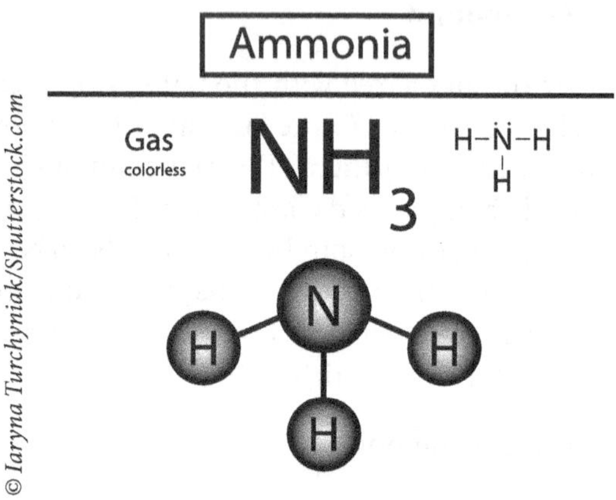

CH_3COOH: 2 C, 2 O, 4 H

Diatomic molecules are compounds that exist naturally are made up of two of the same atoms.

H_2, O_2, N_2, F_2, Cl_2, Br_2, I_2

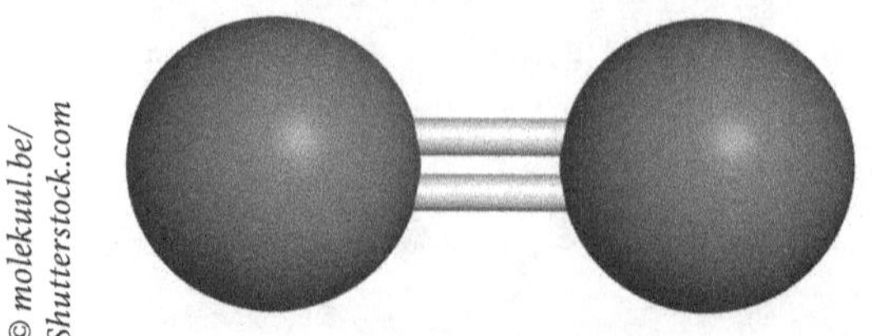

4. **Homogeneous / Pure Substance / Element**

Homogeneous / Pure Substance / Elements are composed of only one kind of atom.

Look on the periodic table for a list of all the elements.

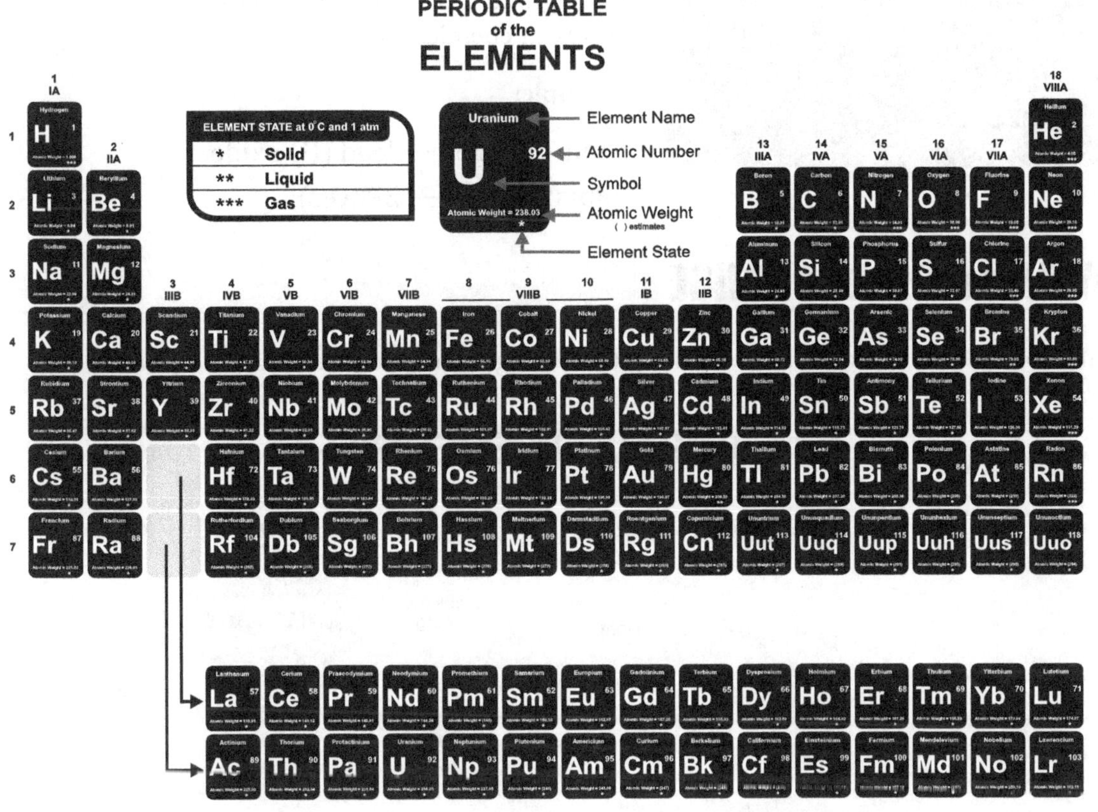

CHEMICAL SYMBOL

A chemical symbol is a shorthand notation to represent an element. Before 1839, most elements were given the neutral (not male or female) ending of Latin *-ium* or Greek *-on*. The element astatine was an exception. Its ending was feminine (*-ine*) and its name derived from the Greek word *astatos*, meaning unstable.

Generally, the first letter of the elemental symbol is capitalized and the other letters are in lowercase. The following are the possible combinations for the symbols:

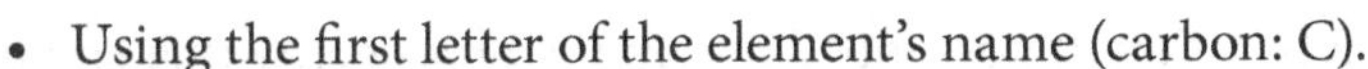

- Using the first letter of the element's name (carbon: C).
- Using the first two letters of the element's name. (silicon: Si)
- Using the first letter and some other letter in the element's name. (magnesium: Mg)
- Using the element's name from another language (Latin or German). The following elements have names that are not common to their chemical symbol.
  - Sb: antimony, Cu: copper, Au: gold, Fe: Iron, Pb: lead, Hg: mercury, Na: sodium, Sn: tin, W: Tungsten, K: potassium, Ag: silver

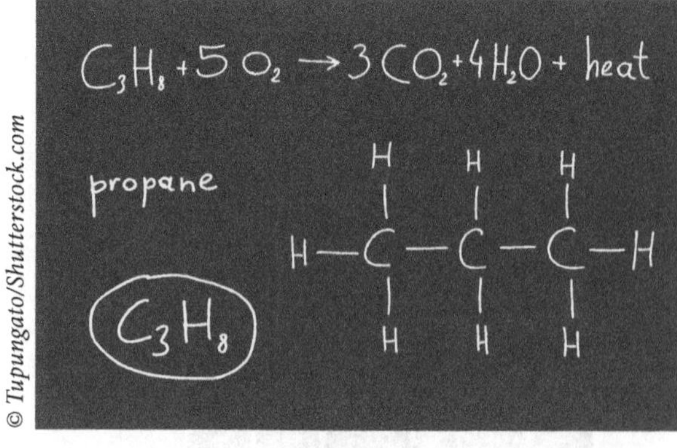

© Tupungato/Shutterstock.com

CHEMICAL EQUATIONS

Lavoisier's discovery of combustion was revolutionary. Other chemists (including Georg Stahl) started to see other chemical changes similar to combustion. These include rusting and respiration. Chemical equations are used to represent chemical changes.

Example:

Lead + oxygen → lead (II) oxide

$2Pb(s) + O_2(g) \rightarrow 2PbO(s)$

ENERGY IN CHEMICAL CHANGE

Energy is the ability to do work. Matter contains stored energy that is released in its movement.

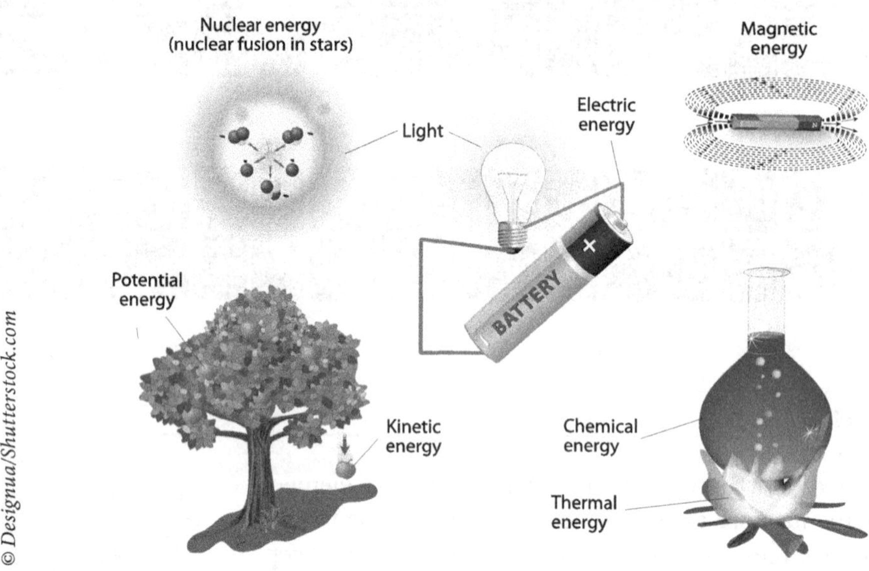

© Designua/Shutterstock.com

Potential vs. Kinetic

- **Kinetic Energy** is energy associated with motion. With matter, temperature is the measurement of kinetic energy. If matter has movement, matter will have a temperature. When matter combines, it absorbs or releases energy in the form of heat, light, or both.
- **Potential Energy** is energy that is stored as a function of position. All matter stores energy that can be released when material reacts in forming or breaking bonds between atoms. We have potential energies that are measured for such bonds.

Endothermic vs. Exothermic

- **Endothermic Reaction** occurs when heat (energy) is absorbed from the surroundings. Energy is positive.
- **Exothermic Reaction** occurs when heat (energy) is released into the surroundings. Energy is negative.

The law of conservation of energy states that energy is conserved; it is neither created nor destroyed (in a non-nuclear change).

Sources:

[1]Bernard Jaffe, *Crucibles: The Story of Chemistry from Ancient Alchemy to Nuclear Fission* (New York: Simon and Schuster, 1984), 50.

[2]http://www.guinnessworldrecords.com/content_pages/record.asp?recordid=57426

[3]http://mooni.fccj.org/~ethall/phlogist/phlogist.htm

[4]Jaffe, *Crucibles*, 50.

Others:

Strathern, Paul. *Mendeleyev's Dream*. New York: St. Martin's Press, 2000.

CHEMISTRY UNIT 1 PRACTICE PROBLEMS

DEFINING MATTER / STATES OF MATTER / PHYSICAL VS. CHEMICAL

Give an Example for Each of the Following:

1. Matter

2. Macroscopic Matter

3. Microscopic Matter

4. Particulate Matter

List the States of Matter, Describe, and Give an Example:

5. ______________________________

Describe:

Example:

6. ______________________________

Describe:

Example:

7. ____

Describe:

Example:

Give Two Examples for Each:

8. A physical property

9. A chemical property

10. A physical change

11. A chemical change

12. List observations that would show the differences between a physical change and a chemical change.

STATES OF MATTER / PHYSICAL VS. CHEMICAL

Fill in the Blanks.

1. ______________ properties can be observed without chemically changing matter.
2. ______________ properties describe how a substance interacts with other substances.
3. ______________ have definite shapes and definite volumes.
4. ______________ have indefinite shapes and definite volumes.
5. ______________ have indefinite shapes and indefinite volumes.
6. ______________ point is the temperature at which a liquid turns to a solid. It is also equal to the ______________ point at which a ______________ turns to a ______________.
7. ______________ point is the temperature at which a gas turns to a liquid.
8. ______________ is the occasional occurrence when a solid turns directly into a gas without turning into a liquid first.

Classify the following as Chemical Change (CC), Chemical Property (CP), Physical Change (PC), or Physical Property (PP).

Blue color	
Flammability	
Dissolves in water	
Scratches glass	
Rusting	
Reacts with oxygen to form a new compound	
Melting point	
Luster (shine)	

Density	
Heaviness	
Boils at 100 degrees	
Sour taste	
Exploding fireworks	
Reacts with H_2O to form gas	
Hardness	
Odor	

Heat conductivity	
Silver tarnishing	
Sublimation	
Magnetizing steel	
Length of metal object	
Melting	
Exploding dynamite	

Combustible	
Water freezing	
Wood burning	
Acid resistance	
Brittleness	
Milk souring	
Baking bread	

Identify the following as being true or false

_____ 1. A change in size or shape is a physical change.
_____ 2. A chemical change means a new substance with new properties was formed.
_____ 3. An example of a chemical change is when water freezes.
_____ 4. When platinum is heated, then cooled to its original state, we say this is a physical change.
_____ 5. When water turns sour, this is a physical change because a change in odor does not indicate a chemical change.
_____ 6. When citric acid and baking soda mix, carbon dioxide is produced and the temperature decreases. This must be a chemical change.

Identify each of the following as a physical or chemical change.

_____ 1. You leave your bicycle out in the rain and it rusts.
_____ 2. A sugar cube dissolves.
_____ 3. Scientists break up water into oxygen and hydrogen gas.
_____ 4. Coal is burned for a barbecue.
_____ 5. A shrub or tree is trimmed because it has grown too tall.

CLASSIFYING MATTER

1. According to the demonstration presented in class, write a chemical reaction and indicate what happens to the mass before and after the reaction.

2. Describe and explain the particulate level of an element compared to a compound.

3. Describe and give an example for each of the following:
   a. Heterogeneous
      a. Describe:

      b. Example:

   b. Homogeneous / Solution
      a. Describe:

      b. Example:

   c. Homogeneous / Pure Substance / Compound
      a. Describe:

      b. Example:

   d. Homogeneous / Pure Substance / Element
      a. Describe:

      b. Example:

4. Fill in the blank
   a. Sb ______________________________
   b. Cu ______________________________
   c. Au ______________________________
   d. Fe ______________________________
   e. Pb ______________________________
   f. Hg ______________________________
   g. Na ______________________________
   h. W ______________________________
   i. K ______________________________
   j. Ag ______________________________
   k. Sn ______________________________

5. List the diatomic molecules

CLASSIFYING MATTER

Classify each of the materials below:

1. Heterogeneous
2. Homogeneous / Solution
3. Homogeneous / Pure Substance / Compound
4. Homogeneous / Pure Substance / Element

Material	1, 2, 3, or 4
Powder laundry detergent (contains white and blue crystals)	
sugar + pure water ($C_{12}H_{22}O_{11}$ + H_2O)	
Iron Filings (Fe)	
limestone ($CaCO_2$)	
orange juice (water and pulp)	
Pacific Ocean (water and salt)	
air	
aluminum (Al)	
magnesium (Mg)	
acetylene (C_2H_2)	
tap water in glass	
pure water (H_2O)	
soil	
chromium (Cr)	
baking soda (NaCl + H_2O)	
salt + pure water (NaCl + H_2O)	
benzene (C_6H_6)	
muddy water	
brass (Cu mixed with Zn)	
Pizza	

Material	1, 2, 3, or 4
Spicy salad dressing	
95% Rubbing alcohol (C_2H_8O + water)	
Captain Morgan (80% Vodka and 20% water)	
compost pile	
gold necklace (gold and copper)	
Smog (dirty air)	
lye (NaOH)	
Nail polish remover (C_2H_6O + water)	
Plaster of Paris (calcium sulfate)	
Opened soda pop	
100% Rubbing alcohol (C_2H_6O)	
Chocolate chip ice cream	
Diamond (carbon)	
Steel (alloy of iron + other elements)	
Black ink (comination of colors)	
Carbon dioxide gas (CO_2)	
Floor tile	
Human tear	
Pre-1982 penny (95% Cu and 5% Zn)	
Blueberry bagel	

ENERGY

Part 1. Definitions of energy.

Write down the definition for each of the following terms:

Energy:

Kinetic energy:

Potential energy:

Part 2. The two basic types of energy.

Determine the best match between basic types of energy and the description provided. Put the correct letter in the blank.

(a) Kinetic Energy
(b) Potential Energy

_____ 1. A skier at the top of the mountain
_____ 2. Gasoline in a storage tank
_____ 3. A race car traveling at its maximum speed
_____ 4. Water flowing from a waterfall before it hits the pond below
_____ 5. A spring in a pinball machine before it is released
_____ 6. Burning a match
_____ 7. A running refrigerator motor

Part 3. Forms of Energy.

Determine the type of energy for each form (Kinetic or Potential).

Form	Definition	Type
Mechanical (motion) energy	An object's movement creates energy	
Thermal (heat) energy	The vibration and movement of molecules	
Electrical energy	Movement of electrons	
Chemical energy	Stored in bonds of atoms and molecules	
Nuclear energy	Stored in the nucleus of an atom; released when nucleus splits or combines	

Endothermic Reactions vs. Exothermic Reactions

Classify each of the following changes as either exothermic or endothermic. Explain your reasoning for each.

Process	Exo	Endo	Explanation
An ice cube melts after being left out on the table.			
Cooking an egg in a frying pan.			
Burning a match.			
The human body uses the energy provided from food digestion.			
Morning dew forming on grass and plants.			
Dynamite explodes in the destruction of a building.			
Making ice cubes.			
A puddle of water evaporates.			
Plants making sugar through photosynthesis.			
Converting frost to water vapor.			
Your own examples			

REVIEW UNIT 1

_____ 1. Which of the following correctly matches the classification as microscopic?
   a. teeth
   b. finger nail
   c. a red blood cell
   d. cat hair
   e. a spoonful of sugar

_____ 2. Which of the following does *not* describe the liquid state?
   a. wet and takes the shape of its container
   b. can measure its volume
   c. can be frozen to a solid
   d. usually has a lower density than a gas
   e. all describe the liquid state

_____ 3. Which of the following describes a chemical change?
   a. dissolving
   b. evaporating
   c. melting
   d. freezing
   e. burning

_____ 4. Which of the following is a solid at room temperature?
   a. mercury
   b. hydrogen
   c. oxygen
   d. neon
   e. silicon

_____ 5. Which of the following is a chemical change?
   a. rusting
   b. melting
   c. freezing
   d. boiling
   e. dissolving

_____ 6. Which of the following changes is/are classified as physical?
   i. carving a block of ice into a sculpture
   ii. iron rusting
   iii. sugar dissolving in water
   iv. A candle burning
   v. souring of cream

   a. i only
   b. v only
   c. i and v
   d. i, iii, and v
   e. i and iii

_____ 7. Breaking water up by separating it into hydrogen and oxygen is an example of a _____.
  a. physical change
  b. chemical change
  c. liquid changed to solid
  d. solid changed to liquid

_____ 8. Which of the following is a chemical property?
  a. combustibility
  b. boiling point
  c. density
  d. odor
  e. taste

_____ 9. Which of the following is a physical property of magnesium?
  a. reacts with bromine
  b. the color is silver
  c. reacts with chlorine
  d. reacts with oxygen
  e. burns easily

_____ 10. Which of the following is classified as a pure substance?
  a. wine
  b. fog
  c. acetone
  d. sand
  e. milk

_____ 11. Classify the material concrete.
  a. heterogeneous
  b. homogeneous / solution
  c. heterogeneous / pure substance / compound
  d. homogeneous / pure substance / compound
  e. homogeneous / pure substance / element

_____ 12. Classify the material 100% hydrogen peroxide.
  a. heterogenous / solution
  b. homogeneous / solution
  c. heterogeneous / compound
  d. homogeneous / compound
  e. heterogeneous / element

_____ 13. Classify the material vodka (alcoholic beverage).
  a. heterogeneous
  b. homogeneous / pure substance / compound
  c. heterogeneous / pure substance / compound
  d. homogeneous / solution
  e. heterogeneous / solution

_____ 14. Which of the following is classified as a pure substance?
  a. sodium
  b. brass
  c. milk
  d. iced tea
  e. soda

_____ 15. Which of the following is a heterogeneous mixture?
a. nitrogen gas
b. filtered air
c. silver
d. alcohol
e. sand

_____ 16. Which of the following correctly describes a homogeneous sample?
a. uniform appearance and composition throughout
b. visibly different parts or phases
c. a mixture of water and oil
d. contains no elements
e. contains no compounds

_____ 17. Which of the following substances is homogeneous / pure substance / compound?
a. a tree branch
b. concrete
c. smog (dirty air)
d. sugar
e. salad dressing

_____ 18. Which of the following substances is heterogeneous?
a. 3% hydrogen peroxide
b. the lint from your clothes dryer
c. a copper wire
d. baking soda
e. iron

_____ 19. What type of substance cannot be decomposed or separated into other stable pure substances?
a. a gas like propane
b. a solution like salt water
c. an element like nickel
d. a compound like CH_4
e. homogeneous matter like 90% rubbing alcohol

_____ 20. Which of the following is *not* a diatomic molecule?
a. hydrogen
b. oxygen
c. chlorine
d. sulfur
e. nitrogen

_____ 21. Which of the following is an exothermic changes.
a. a cold pack
b. a hot pack

_____ 22. When paper is burned, the mass of the remaining ash is less than the mass of the original paper. Does the behavior happen with all material when it burns?
a. yes
b. no

HISTORY

1. What are the four elements in Aristotle's theory of matter?

2. What is the philosopher's stone and how does it relate to early chemistry?

3. Define alchemy?

4. Where did many of our earliest understandings of chemistry evolve?

5. What is phlogiston? Does it exist?

6. Who disproved the phlogiston theory and gave oxygen its name?

7. According to the phlogiston theory, what is the carrier of phlogiston?

MATTER

1. Define matter.

2. Discuss how a sample of sugar could be classified as:
   a. Macroscopic material

   b. Microscopic material

   c. Particulate material

3. Who proposed the four elements of matter are earth, water, fire, and air?

4. Classify the following as macroscopic, microscopic, or particulate: a carbon dioxide molecule.

5. Classify the following as macroscopic, microscopic, or particulate: a red blood cell.

6. The word pour is commonly used in reference to liquids, but not to solids or gases. Can a solid or gas be poured? Why or why not? If either answer is yes, can you give an example?

7. Is it possible to melt a piece of ice without seeing a change in temperature? Explain.

8. Which state of matter (solid, liquid, or gas) has no atoms moving?

PHYSICAL AND CHEMICAL PROPERTIES AND CHANGES

1. Classify the following as chemical or physical properties.
    a. the mass of a chunk of nickel is 19 grams
    b. when sulfur is exposed to water it burns
    c. the color of copper is a rust-like color
    d. sugar can dissolve in water
2. Classify the following changes as chemical or physical.
    a. rusting car
    b. boiling water
    c. digesting rice in the stomach
    d. dropping Alka Seltzer™ in water
3. Which of the following is an example of a chemical change?
    a. water boiling
    b. ice melting
    c. natural gas burning
    d. iodine vaporizing
    e. dry ice turning from the solid phase to the gas phase
4. List some chemical properties of water?

CLASSIFYING MATTER

1. The smallest particle of an element that retains the chemical properties of the element is a(n)?

2. To what category would a sample containing more than one phase belong?
   a. compound
   b. element
   c. diatomic mixture
   d. homogeneous mixture
   e. heterogeneous mixture

3. Which of the following is a homogeneous solution?
   a. vinegar (5% acetic acid)
   b. water
   c. baking soda
   d. sewage
   e. diamond

4. Classify boron, carbon dioxide, methane, and lead as element or compound.

5. What is the symbol for Antimony?

6. List the diatomic molecules.

7. Which of the following is a pure substance?
   a. vinegar (5% acetic acid)
   b. concrete
   c. baking soda
   d. sewage
   e. brass

8. How would you classify steel?
   a. heterogeneous mixture
   b. homogeneous, solution
   c. homogeneous, compound
   d. homogeneous, element

9. What is the chemical symbol for mercury?

10. Which of the following is a physical combination of two or more pure substances?
    a. salt
    b. air
    c. sand
    d. water
    e. natural gas

11. What is a mixture of 80% nitrogen (N_2) and 20% oxygen (O_2) which can be separated into these compounds by cooling (physical change) is commonly called?

12. What does the following chemical formula indicate? Al_2O_3
    a. 3 atoms aluminum, 2 atoms oxygen
    b. 1 atom aluminum, 1 atom oxygen
    c. 1 atom aluminum oxygen
    d. 6 atoms aluminum oxygen
    e. 2 atoms aluminum, 3 atoms oxygen

13. Which of the following is a chemical compound?
    a. Co
    b. CO
    c. Pb
    d. Sn
    e. Ar

14. Which of the following is a chemical compound?
    a. alcohol
    b. aluminum
    c. antimony
    d. argon
    e. astatine

15. How would you classify soda pop?
    a. heterogeneous mixture
    b. homogeneous, solution (unopened soda pop)
    c. homogeneous, compound
    d. homogeneous, element

16. How would you classify propane?
    a. heterogeneous
    b. homogeneous, solution
    c. homogeneous, pure substance

17. Which of the following is a diatomic molecule?
    a. Ar_2
    b. Br_2
    c. Cr_2
    d. Dr_2
    e. Er_2

18. Write the chemical name or symbol for each of the following:

Krypton	Al	Chromium
Carbon	Cl	Tellurium
Mercury	P	Protactinium

19. Indicate which category or categories each of the following substances belongs.

Substance	Heterogeneous	Homogeneous	Solution	Pure Substance	Element	Compound
Pure NaCl						
Al foil						
Filtered air						
Concrete						
Sand						
Plastic (polystyrene)						
Astatine						
Salt water						

CHEMICAL REACTIONS & ENERGY

1. Write the chemical equation for the following described reactions. Identify the reactants and products (It is alright if your equation is not balanced, we will address this topic in a future unit).
   a. Propane (C_3H_8) burns (in the presence of oxygen) to produce carbon dioxide (CO_2) and water.
   b. Energy is required to separate NaCl into its elements.
2. Classify the following as kinetic or potential energy.
   a. a running halfback
   b. a bowling ball at the top of the stairs
   c. water flowing over the falls
   d. a moving train
   e. a strap holding down a bike on a trailer
3. Classify the following as an endothermic or exothermic reaction.
   a. ice melting
   b. wood burning
   c. an ice pack (for sports injuries)
   d. a heating pad

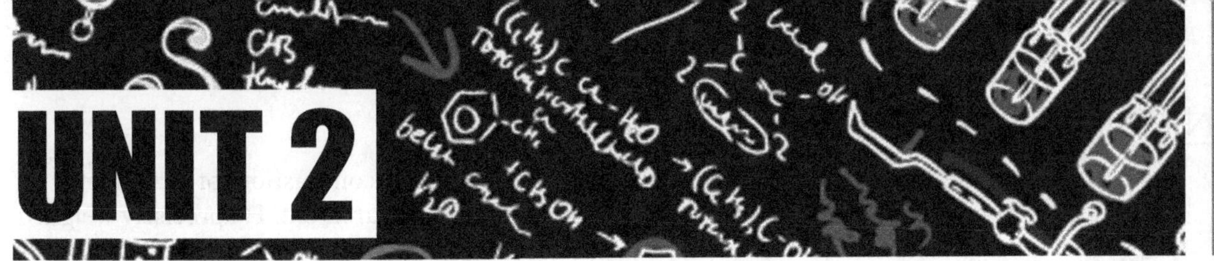

MEASURING MATTER (MEASUREMENT)

OBJECTIVES FOR UNIT 2

- To learn the history of measurement.
- To learn how to write a number in scientific notation.
- To know the different metric units and conversions for length, mass, time, temperature, volume, and density and their interrelationships.
- To know the unit values in the metric system.
- To understand the difference between precision and accuracy.
- To determine the number of significant figures in a number and to be able to use significant figures correctly in calculations.
- To learn the measurement and calculation of volume and density.

MEASUREMENT TIMELINE

0

measurement gains popularity
1800s

1875
Treaty of the Meter signed

SI (metric system)
1960

WHEN DID WE START USING MEASUREMENT?

What made Lavoisier's experiment on combustion of lead metal so unique was the fact that it involved measurement. He precisely weighed the samples and equipment both before and after the experiment. This was not very common in that era for science. A few others such as Becher, Boyle, and Cavendish also used measurements, but measuring certainly wasn't at the forefront of their procedures. After all, the four element theory was the truth of the time so what was the necessity of measuring? Science just involved a recombination of the same four elements, so why keep track of their weights? The bakers and tax collectors of the time were more analytical than people working in the laboratory.

Lavoisier presented the benefit of measurement. As others began to see measurement as essential in the discovery of more elements and compounds, science started to move ten-fold. In the early 1800s, Jöns Jakob Berzelius, a Swedish chemist, performed some of the first measured chemical experiments with the electric battery. At this time 49 elements had been discovered and classified. A famous English chemist named John Dalton also used measurement to prove the existence of atoms. (We will discuss this later.)

So, what kind of measurements are we talking about? Well, weighing was already mentioned with Lavoisier's experiment. There were also measurements of how much space matter took up (we call this "volume").

THE EVOLUTION OF UNITS

What about the units, how were measurement recorded?

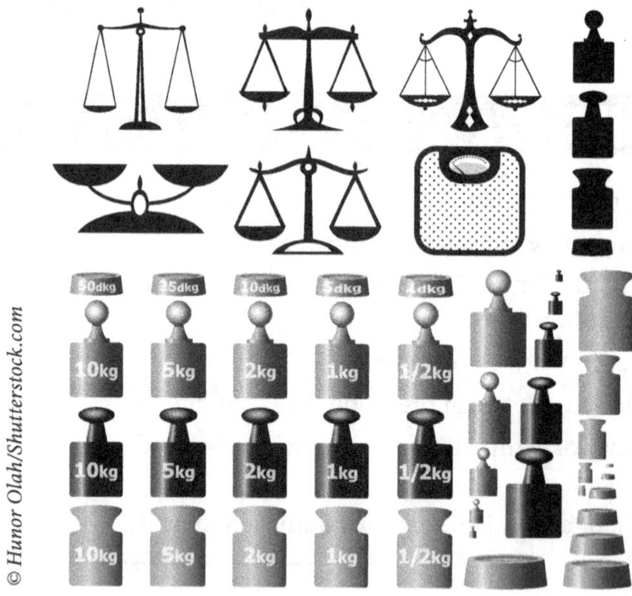

Before the introduction of the metric system at the end of the 18th Century, a bewildering array of measures was used. Some countries, such as France, had up to 800 different measures by name. More confusing still, even measures with the same name could describe very different quantities. The tiny duchy of Baden in Germany had 112 different ells (a unit of measure roughly corresponding to the length of a man's arm from the elbow).

Non-metric measures suffered from the following inconsistencies:

- Measures varied by region (a London pound was different from a Newcastle pound)
- Measures varied by object measured (a gallon of beer was different from a gallon of wine)
- Measures varied by trade or context (the nautical mile was different from the geographical mile)
- Measures varied by the origin of the measure (Britain used at least three different sets of measures: avoirdupois, troy, and apothecary)
- Measures changed over time (The pint of 1850 was different from the pint of 1800)

This chaos was the norm in most countries.[1]

METRIC SYSTEM (SI—INTERNATIONAL SYSTEM): STANDARD UNIT OF MEASURE

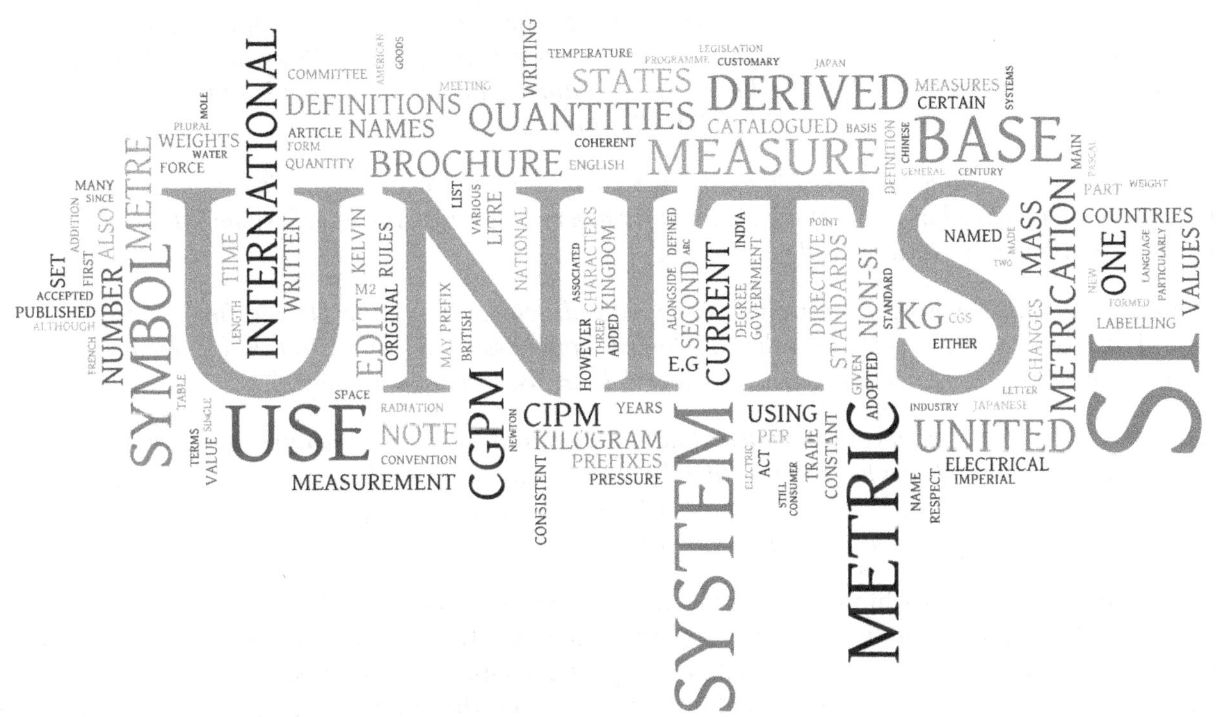

© goodcat/Shutterstock.com

All measurements contain at least two elements: a number which indicates the magnitude of the quantity being measured, and the units which provide the basis for comparing the measured quantity to a known reference. There are several different systems of units, each of which contains units of measurement for properties such as distance, volume, and weight. In the British system, now used in the United States (and almost nowhere else), the various units of measurement are defined in a purely arbitrary manner. There are 12 inches in one foot, but three feet in one yard. There are 16 ounces (by weight) in one pound, and 32 ounces (by volume) in one quart. There are two pints to a quart, and four quarts to a gallon, except in the British Empire where a gallon contains five quarts and is called an imperial gallon.

According to Madsci Network:

The metric system, starting in France during the 1790s had enjoyed considerable success and spread to many European countries throughout the 1800s. During the 1870s, the French government decided to make the metric system truly international by installing an international body to oversee weights and measures.

© Jojje/Shutterstock.com

On 20 May 1875, the Treaty of the Meter was signed by 17 countries. This established the International Bureau of Weights and Measures (BIPM), which is still in charge of our weights and measures. It receives its instructions from the International Committee for Weights and Measures (CIPM), which in turn is authorized by the General Conference on Weights and Measures (CGPM). Delegates from all member states of the Treaty of the Meter attend the CGPM meetings. At present, the CGPM meets every four years—this is when major decisions affecting our weights and measures are being taken. This international and democratic nature of the metric system led to it being adopted in all countries.

Some countries (including the U.S.) still permit the use of some non-metric units, but even these units have been defined in terms of metric units for a very long time. The metric system thus is the basis for all measuring worldwide.

The metric system is not static; it evolves along with all science. Since the 11th meeting of the CGPM in 1960, the metric system is called SI (International System of Units). In this document, the terms metric system, modern metric system and SI are used interchangeably.

Evolution of the metric system does not mean that the magnitudes of units are changing. A metric unit of a certain name always has meant and will mean exactly the same thing, no matter where, or when, or how used. This is one of the great achievements of metric: stability, reliability and reduction of error, confusion and misunderstandings.[1]

One of the fundamental ideas behind the metric system was the use of only one measure per physically measurable quantity. To avoid having to use very small or large numbers with these measures, prefixes were invented. Prefixes (such as *milli-* or *kilo-*) are qualifiers that change the magnitude of a measurement unit when put in front of the unit name. By using this system of units and prefixes, the number of units with different names can be greatly reduced—weights and measures become simpler.

The prefixes were designed to work nicely with the decimal system; that is our system of writing numbers. The decimal system had been used for thousands of years and was the only numbering system in use in recent history. Prefixes can be changed by simply moving the decimal marker—no conversions are necessary.

Another idea was to define every measurement unit in terms of base units. The metric system today has seven such base units; all other units can be converted into base units. For these unit conversions, conversion factors are not needed. The conversion factor is usually 1, thus eliminating mistakes and inaccuracy through rounding.

The metric system created, for the first time, a system that allowed weights and measures to be understood by anybody.

SI BASE UNITS

Physical Quantity	Name of Unit	Symbol
length	meter	m
mass	kilogram	kg
time	second	s
electric current	ampere	A
temperature	Kelvin	K
luminous intensity	candela	cd
amount of substance	mole	mol

DIFFERENT MEASUREMENTS

1. **Weight and mass**

Weight is a measure of the force of gravity between two objects. Force changes as the distance from the object to the earth (two objects) changes.

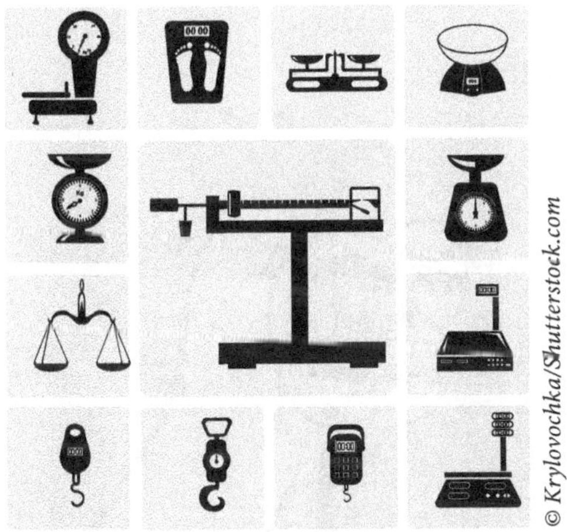

- Unit for weight: pounds
  1 lb = 16 oz
  1 ton = 2,000 lbs
- Weight is measured with a scale.

Mass is used because it's a measurement that doesn't change from place to place. It's a measure of the quantity of matter.

- Unit for mass: kg or (grams)
  1 kg = 1,000 g
- Mass is measured with a balance. This instrument is used to determine the mass of an object by comparing the unknown mass to a known "prototype mass" (the kilogram). This process is called massing, not weighing. (These terms are often misused.)

2. **Length**

Length is the distance covered by a line segment connecting two points.

© LesPalenik/Shutterstock.com

- Unit for length: meter
  1 inch = 2.54 cm
  1 km = 0.6237 mile
- Length is measured with a meter stick.

3. **Time**

Time is the interval between two occurrences.

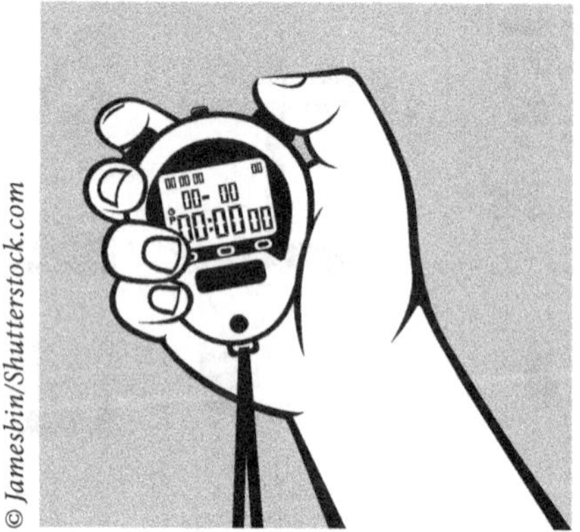
© Jamesbin/Shutterstock.com

- Unit for time: second
  3600 seconds = 1 hour
- Time is measured with a stopwatch.

4. **Temperature**

Temperature is a measure of the average kinetic energy of the particles that make up the sample.

A picture of Lord Kelvin (William Thomson).

© Nicku/Shutterstock.com

- SI unit for temperature is: K (Kelvin).
  Kelvin is related to Celsius degree by the following equation:

  K = °C + 273.15 or °C = K – 273.15

  Relating °C and °F

  °F = (1.8 × °C) + 32

- Temperature is measured with a thermometer.

Anders Celsius who is contributed to developing the Celsius scale based it on two fixed temperatures: the temperature of thawing snow and ice (0°C) and the temperature of boiling water (100°C).

© Olga Popova/Shutterstock.com

Melting point is the temperature at which a solid turns into a liquid.

Boiling point is the temperature at which a liquid turns into a gas.

Absolute zero is the lowest possible temperature (zero on the Kelvin scale).

Examples of calculating temperature:

© petr73/Shutterstock.com

The temperature today is 77.0°F. Calculate this temperature in °C and K.

°C = (°F – 32)/ 1.8

°C = (77.0 – 32)/ 1.8 = 25.0°C

SCIENTIFIC NOTATION

To handle either very large or very small numbers, we use a system called scientific notation.

In one gram of the element carbon there are 50,150,000,000,000,000,000,000 atoms of carbon. The mass of a single carbon atom is 0.000000000000000000000019924 grams. Imagine how cumbersome it would be if you wanted to find the mass of 2150 carbon atoms. Using scientific notation will greatly simplify the problem.

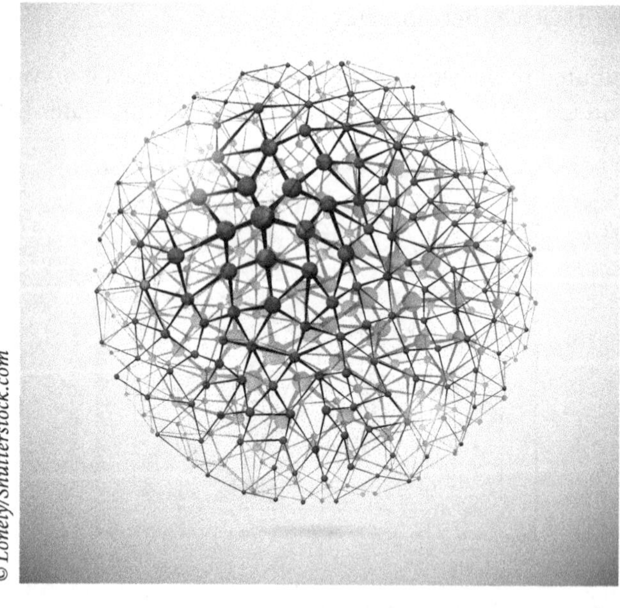
© Lonely/Shutterstock.com

To express a number in scientific notation, write it in the form:

$N \times 10^{n}$

N = a number between 1 and 10.

n = the number of places to get N between 1 and 10.

Examples:

$538{,}000{,}000 = 5.38 \times 10^{8}$

$0.00045 = 4.5 \times 10^{-4}$

If a number is larger than 1:

$N \times 10^{n}$

- The exponent (n) is the number of places you move the decimal point.
- The original decimal point is moved (n) number of places to the left.
- The resulting number is multiplied by 10^{n}.

Large number examples:

$60{,}230{,}000{,}000{,}000{,}000 = 6.023 \times 10^{16}$

$214{,}570{,}000{,}000 = 2.1457 \times 10^{11}$

$31.47 = 3.147 \times 10^{1}$

If a number is smaller than 1:

$N \times 10^{-n}$

- The exponent (n) is the number of places you move the decimal point.
- The original decimal point is moved (n) number of places to the right.
- The resulting number is multiplied by 10^{-n}.

Small number examples:

$0.000040 = 4.0 \times 10^{-5}$

$0.000000000000326 = 3.26 \times 10^{-13}$

$0.000458000003 = 4.58000003 \times 10^{-4}$

METRIC CONVERSIONS

SI PREFIXES

The SI system also permits the use of certain prefixes which define decimal fractions and multiples of the base SI units. Some of the SI prefixes are shown in the table below.

Prefix	Abbreviation	Scientific Notation	Example
mega-	M	10^6 (one million)	1 Mm = 1,000,000 m
kilo-	k	10^3 (one thousand)	1 km = 1,000 m
hecto-	h	10^2 (one hundred)	1 hm = 100 m
deca-	da	10^1 (ten)	1 dam = 10 m
		10^0 or 1 (one)	1 meter
deci-	d	10^{-1} (one tenth)	10 dm = 1 m
centi-	c	10^{-2} (one hundredth)	100 cm = 1 m
milli-	m	10^{-3} (one thousandth)	1,000 mm = 1 m
micro-	μ	10^{-6} (one millionth)	1,000,000 μm = 1 m
nano-	n	10^{-9} (one billionth)	1,000,000,000 nm = 1m

METRIC CONVERSION PROBLEMS

Because of the global world we live in, we are constantly encountering conversion problems. This may be as simple as determining how many miles you run in a 5-kilometer run. Look on any package at a grocery store; you will see units of grams, ounces, liters, and many more. So how much food actually is in 30 grams or 240 milliliters size servings?

In chemistry, we often deal with determining mass, temperature, and volume of substances to study their behavior. Understanding how to convert between units (mostly in the SI system) is a necessity!

There are two approaches to metric conversion problems:

Approach 1: Dimensional Analysis (Factor-Label Method)

1. Identify what units the final answer should be.
2. Use the given measurements as a starting point.
3. Write down the connecting conversion factors.
4. Perform the calculation.
5. Check answer for sensibility.

Examples:

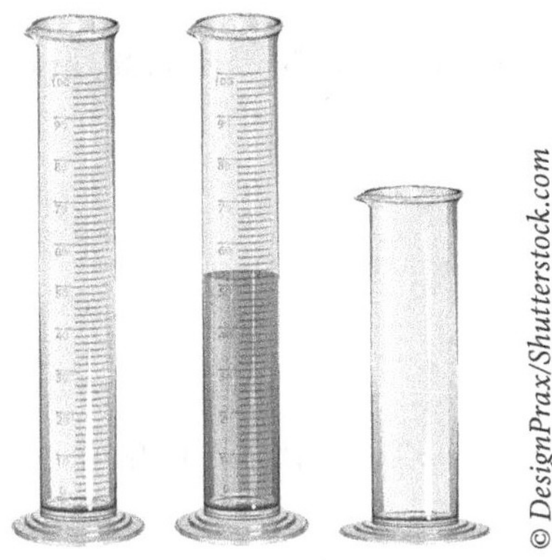

© DesignPrax/Shutterstock.com

What is the volume (in mL) of 0.354 liters of vinegar?

$$0.345L \times \frac{1000mL}{1L} = 345mL$$

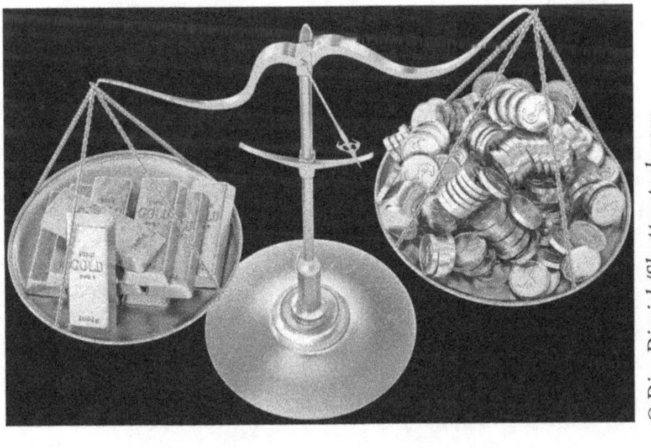

© Dim Dimich/Shutterstock.com

How many kilograms are there in 89 mg of gold?

$$89mg \times \frac{1g}{1000mg} \times \frac{1kg}{1000g} = 0.000089kg$$

A solid cube of copper metal is 120.00 dm long, 8.50 cm wide, and 5.00 mm thick. Calculate the volume of copper.

$$120dm \times \frac{1m}{10dm} \times \frac{100cm}{1m} = 1200cm$$

$$5.00mm \times \frac{1m}{1000mm} \times \frac{100cm}{1m} = 0.500cm$$

$$volume = 1200cm \times 8.50cm \times 0.500cm = 5100cm^3 = 5100mL$$

Approach 2: Stairstep Method

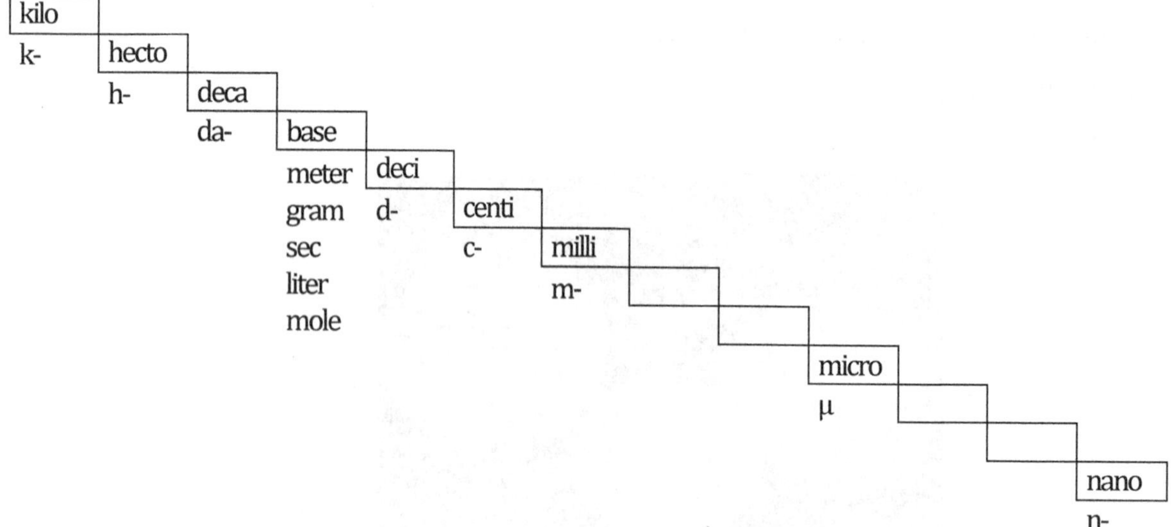

Examples:

807 ds = ? ms (move 2 "down the steps" move decimal to the right 2 steps)
80700 ms

0.00579 ng = ? cg (move 7 "up the steps" move decimal to the left 7 steps)
0.000000000579 cg = 5.79×10^{-10} cg

PRECISION VS. ACCURACY

Precision indicates how close together or how repeatable the results are. A precise measuring instrument will give very nearly the same result each time it is used.

More precise		Less precise	
Trial #	Mass (g)	Trial #	Mass (g)
1	100.00	1	102.10
2	100.01	2	100.00
3	99.99	3	99.88
4	99.99	4	98.02

Accuracy indicates how close a measurement is to the accepted value. For example, we'd expect a balance to read 100 grams if we placed a standard 100 g weight on the balance. If it does not, then the balance is inaccurate.

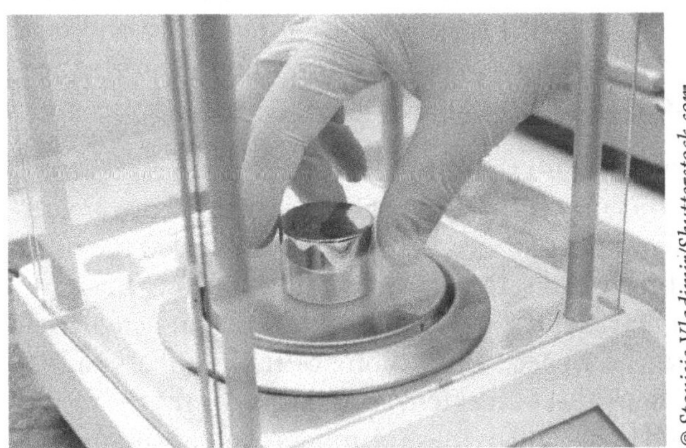

© Stanisic Vladimir/Shutterstock.com

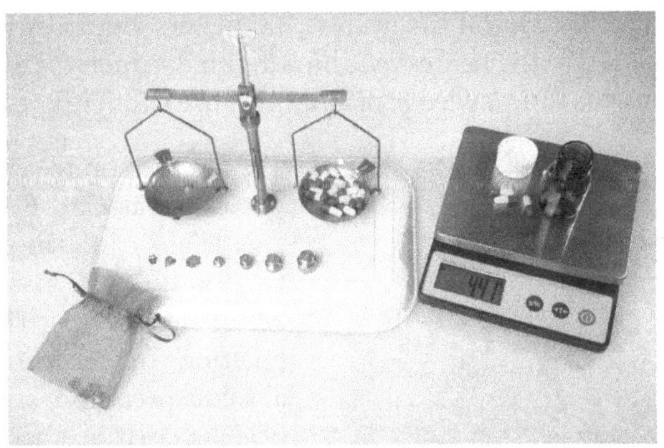

© smoxx/Shutterstock.com

SIGNIFICANT FIGURES

HISTORY OF SIGNIFICANT FIGURES

© B. Melo/Shutterstock.com

Take a Roman numeral like CCLXX. One thing about numbers written in this form is that every C, L, and X counts towards something. Put another way, every digit signifies something. In our modern, Arabic numerals, this would be written 270. The 0 at the end doesn't add value to the number; it is just a place keeper increasing the value of the non-zero digits to the left of it. Before the introduction of zero, every digit was a significant digit; it was when zeros came on the scene that this began to change.

The earliest date the *Oxford English Dictionary (2nd edition)* records anyone using significant figures to mean a nonzero number is around 1400, about two to three hundred years after the time that Arabic numerals entered Europe (see http://www.britannica.com. Search for "Arabic numerals"). So the earliest use of the times just meant those figures to the left of the right-most zeros.

As we move towards modern times, the idea of significant figures (sig figs) took on new meaning. It became a quick way to indicate the precision of a measurement. If I write 1800 meters, then I have only measured the distance to around the nearest 100 meters, but if I write 1837.35, then I have 6 sig figs of precision and have measured to the nearest one hundredth of a meter. This is an example of round-off error, where errors in measurements or knowledge of a number are explicitly recorded via significant figures.

© Nicku/Shutterstock.com

The first person to seriously deal with round-off error and its effects on computation was Carl Friedrich Gauss (1777–1855). One of the top three mathematicians of all time (Euler and Archimedes are usually accorded the other two top spots), Gauss was the first to seriously look at how limiting the number of sig figs that trigonometric and logarithmic tables used affected computation. In *Theoria Motus*, he writes, "Since none of the numbers...admit absolute precision, but are all to a certain extent approximate only, the results of all calculations performed by the aid of these numbers can only be approximately true." Gauss went on to put his money where his mouth was and work out the mathematics of error analysis, a form of which appears as our rules for determining the number of sig figs after a multiplication, addition, or square root operation.[2]

USING SIGNIFICANT FIGURES FOR PRECISION

When we collect data we record our numbers with a certain amount of digits in specific place values. The place values recorded indicates the precision of the instruments used.

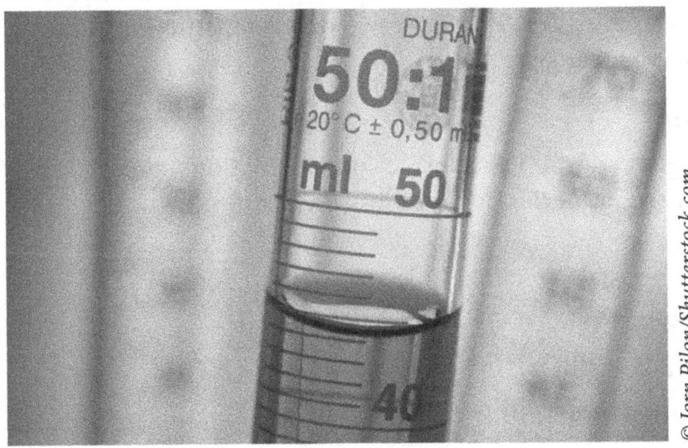

© Jorn Pilon/Shutterstock.com

All numeric data recorded in the laboratory must show the precision of the instruments. We call this recorded numeric data as being significant when precision is shown.

Example:

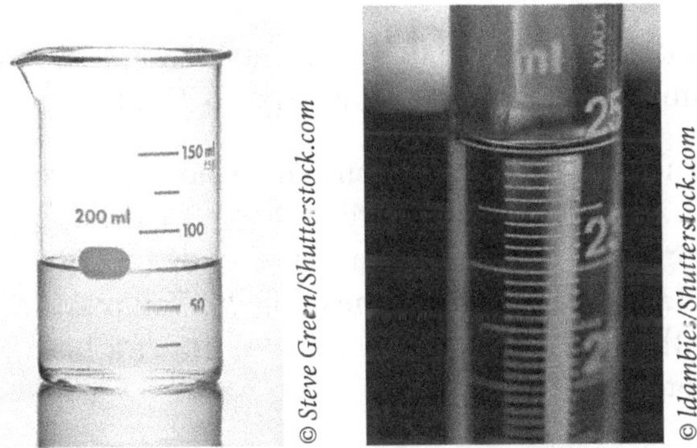

© Steve Green/Shutterstock.com

© ldambies/Shutterstock.com

Three volumes were taken. The numeric data recorded is as follows.

Instrument	Precision	Recorded Number
beaker	±1 mL	46 mL
graduated cylinder	±0.1 mL	44.6 mL
pipet	±0.01 mL	45.00 mL

To average this data:

- 46 mL + 44.6 mL + 45.00 mL = 135.6 mL
- 135.6 mL ÷ 3 = 45.2 mL
- For showing precision, the answer = 45 mL

The answer to the calculated average must show the precision from the instruments used in the experiment. Therefore, the answer is based upon the least precise instrument (beaker).

RULES FOR SIGNIFICANT FIGURES

A lot of numeric data collected in the laboratory will undergo calculations. These calculations may involve addition, subtraction, multiplication, and/or division. As seen in the above example, the numeric answer recorded reflects the precision of the instruments used. In other words, the numeric answer must contain the correct amount of significant answer. There are two different situations that involve recording the correct amount of significant figures in an answer.

1. **Addition and Subtraction**

If your numeric data undergoes addition or subtraction calculations, precision is kept by recording the final numeric answer based upon the numeric data with less precision shown.

Examples:

140 g shows less precision than 141 g

140 g + 141 g = ~~282 g~~ = 280 g

4.9 L shows less precision than 4.89 L

4.89 L + 4.9 L = ~~9.79 L~~ = 9.8 L

340,000 m shows less precision than 234,000 m

340,000 m – 234,000 m = ~~106,000 m~~ = 110,000 m

456.3 s shows less precision than 23.33 s

456.3 s – 23.33 s = ~~432.97 s~~ = 433.0 s

Now, you may be asking, "Isn't it more accurate to record the last answer as 432.97 s?" It would if your data allowed such precision. The number 456.3 s limits the answer to 433.0 s.

How about this example. You and a friend both run one lap around a track. You are both using two different stopwatches. Your stopwatch is more precise and it records your time at 68.20 seconds. Your friend's stop-watch only has enough precision to record her time at 70 seconds (not 70. Seconds; yes, that decimal says a lot!). Her time could be 66.44, 73.03, or even your time of 68.20. You just don't know her time as precisely as you know yours. Now, let's say that you wanted to find a total time for both you and your friend. Could you see with great precision what your total time was?

????? 68.20 s + 70 s = 138.20 s ????

Of course not, because that calculation would indicate that you knew that your friend ran exactly 70 seconds.

Therefore, the correct answer is 140 seconds.

2. **Multiplication and Division**

If your numeric data undergoes multiplication or division calculations, precision is also kept by recording the final numeric answer based upon the numeric data with less precision shown. But, because the decimal point moves during the multiplication and division calculation, it can be difficult determining precision for your final answer. With multiplication and division, we use a method called counting significant figures, which helps us determine how our final answer is recorded.

Rules for counting significant figures:

1. All nonzero numbers are significant.

   458,999—six sig figs

   3.45 x 10^2—three sig figs

2. Leading zeros are not significant
   0.421—three sig figs
   0.0045—two sig figs
3. Captive zeros (zeros between significant figures) are significant
   4012—four sig figs
   9.05 × 107—three sig figs
4. Trailing zeros are significant if and only if a decimal point is shown.
   114.20—five sig figs
   5.90000 × 10^{-4}—six sig figs

Now that you know how to count significant figures, finding the answer is easy. For multiplication or division, follow these steps:

1. Determine the number of significant figures for each number in the calculation.
   6.5 cm × 98.0 cm × 0.04389 cm = ?
   6.5 cm—two sig figs, 98.0 cm—three sig figs, 0.04389 cm—four sig figs
2. Perform the multiplication or division, always following the order of operation.
   6.5 cm × 98.0 cm × 0.04389 cm = 279.5793 cm^3
3. Record your answer with the least amount of significant figures shown in the beginning numbers.
   279.5793 cm^3 = 280 cm^3 (6.5 only had two sig figs, the answer is recorded with only two sig figs)

Examples:

$$\frac{\left(1.490\times10^{33}\right)\times\left(4.70032\times10^{-9}\right)}{\left(3.0079\times10^{-4}\right)}=2.328\times10^{28}$$

$$\frac{1.43\times2.658}{\left(2.65+0.01\right)}=1.43$$

DERIVED UNITS: VOLUME AND DENSITY

Derived units are units that are created by combining base units.

Quantity	Dimensions	SI Units	Common Name
Volume	length × length × length	cm^3	milliliter
Area	length × length	m^2	square meter
Velocity	length/time	m/s	
Density	mass/volume	kg/m^3	
Frequency	cycles/time	s^{-1}	hertz (Hz)
Acceleration	velocity/time	m/s^2	
Force	mass × acceleration	kg m/s^2	Newton (N)
work, energy, heat	force × distance	kg m^2/s^2	Joule (J)

1. **Volume**

Volume: When matter takes up space, it occupies volume.

R. Gino Santa Maria/Shutterstock.com

- SI unit for volume: liter
  $1\ L = 1000\ mL = 1000\ cm^3 = 1\ dm^3$
  $1\ mL = 1\ cm^3$
- Volume is measured with a graduated cylinder (or calculated taking length measurements).

2. **Density**

Density is the measurement of the mass of an object, divided by the volume of the object.

One "common sense" way to look at density is: The more mass that is packed into a volume, the higher the density.

© Peter Hermes Furian/Shutterstock.com

$$Density = \left(\frac{mass}{volume}\right)$$

$$D = \frac{m}{V}$$

- SI unit for density: uses mass divided by volume g/cm^3, g/mL, or g/L, English densities are lbs/ft^3 or lb/gal.
- Density is calculated from a measured mass and volume.

Matter	Density (g/cm^3)	Matter	Density (g/cm^3)
Air	0.0013	Bone	1.7–2.0
Water	1.0	Urine	1.01–1.03
Gold	19.3	Gasoline	0.66–0.69

Example Calculation:

Water was placed into a graduated cylinder until the volume read 25.0 mL. An irregularly shaped piece of metal with a mass of 50.8 grams was placed in the cylinder and completely submerged. The water level rose to the 36.2 mL mark on the graduated cylinder. What is the density?

$$Density = \left(\frac{50.8g}{(36.2-25.0)mL}\right)$$

$$D = \frac{m}{V} = 4.54g/mL$$

Sources:

[1]http://homepage.ntlworld.com/cdkaese/home.htm

[2]http://www.madsci.org/posts/archives/jan2000/947992509.Sh.r.html

Others:

http://antoine.frostburg.edu

http://www.carlton.paschools.pa.sk.ca/chemical/Sigfigs/accuracy_and_precision.htm

CHEMISTRY UNIT 2 PRACTICE PROBLEMS

MEASUREMENT / SCIENTIFIC NOTATION / METRIC CONVERSIONS

Based upon the following pictures:

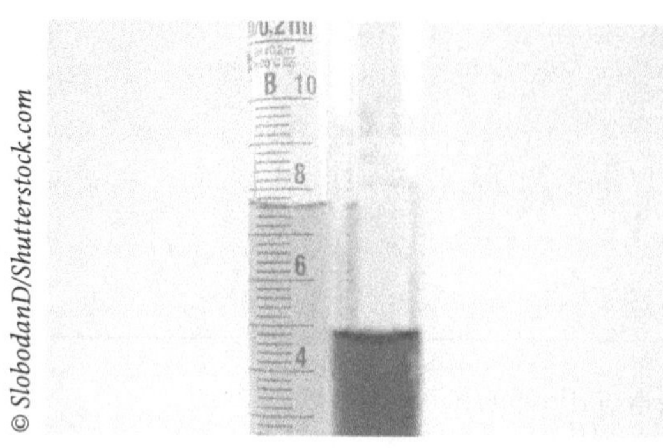
© SlobodanD/Shutterstock.com

1. How would you record the volume of this liquid? ____________

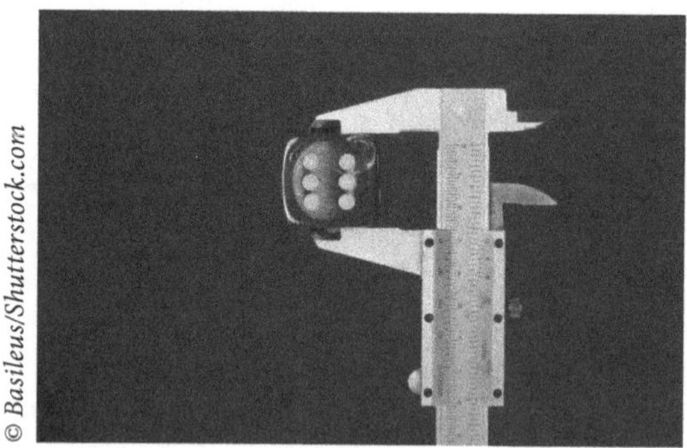
© Basileus/Shutterstock.com

2. How would you record the length of this die? ____________

3. Write the following numbers in scientific notation:
   a. 321,012,000 a. ____________
   b. 0.0000330000 b. ____________
   c. 0.000098500 c. ____________
   d. 4.67 d. ____________

4. Write the following numbers in ordinary decimal notation:
   a. 4.32×10^{6} a. ____________
   b. 4.32×10^{-2} b. ____________
   c. 9.800×10^{-7} c. ____________
   d. 4.006000×10^{9} d. ____________

5. Convert the following units:

a. 5.68 kg to mg	a. ______________
b. 0.000987 dm to mm	b. ______________
c. 8.76×10^{4} mL to L	c. ______________
d. 5.4×10^{-4} μs to ms	d. ______________
e. 9.56×10^{9} dL to kL	e. ______________
f. 78.998 mg to ng	f. ______________
g. 0.0087 ns to cs	g. ______________
h. 8.99×10^{-4} cm^3 to μL	h. ______________

Write the following number into scientific notation:

1. 100,000,000,000	2. 0.0034
3. 699,000,000	4. 0.0000597
5. 3,003	6. 0.00908070

Write the following number into ordinary decimal notation:

7. 1.0×10^{8}	8. 2.3×10^{3}
9. 77×10^{5}	10. 8.76×10^{-4}
11. 9.3×10^{-3}	

Convert the following:

12. 10 km to cm	13. 1×10^{6} mL to L
14. 1×10^{9} kg to μg	15. 100 m to km
16. 1×10^{4} ng to mg	17. 6.5 m to dm
18. 4.56 dg to mg	19. 0.689 ks to cs
20. 0.00223 μL to mL	21. 5,555,000 nm to dm
22. 3.58 ms to ns	

Determine the number of significant digits in each of the following:

23. 5.40	24. 1.2×10^{3}
25. 210	26. 0.00120
27. 801.5	28. 0.0102
29. 1,000	30. 9.010×10^{-6}
31. 101.0100	32. 2,370.0

Add:

33. 16.5 + 8 + 4.37	34. 13.25 + 10.00 + 9.6
35. 2.36 + 3.38 + 0.355 + 1.06	36. 0.0853 + 0.0547 + 0.0370 + 0.00387
37. 25.37 + 6.850 + 15.07 + 8.056	38. 2390 + 123.9

Subtract:

39. 23.27 – 12.058	40. 13.57 – 6.3
41. 27.68 – 14.369	42. 350.0 – 200

Multiply:

43. 2.6 × 3.78	44. 3.08 × 5.2
45. 6.54 × 0.37	46. 0.0036 × 0.02

Divide:

47. 45.00/8.9	48. 0.09/0.31
49. 8.6700/0.002	50. 34.005/0.0098

MEASUREMENT / SCIENTIFIC NOTATION / METRIC CONVERSIONS

1. Describe how each is measured and give a SI unit for each.
   a. mass
   b. length
   c. time
   d. temperature
   e. volume

2. Convert the following units:
   a. 4.56 dg to mg
   b. 0.00223 microliters to mL
   c. 0.689 ks to cs
   d. 5,555,000 nm to dm

3. Determine the number of significant figures in each of the following numbers:
   a. 54.0003 mL
   b. 0.0003200 cm^3
   c. 34.003 mL
   d. 45,500 cm

4. Round the following numbers to three significant figures:
   a. 34,690
   b. 0.4500000
   c. 210,000

5. Calculate the following and record your answer with the proper number of significant figures:
   a. 54.900 + 32.1
   b. (65.00 × 0.21) / 0.03
   c. 43,000 × 65.00
   d. (34.56 – 34.01) × (23.4 + 11.1)

6. Convert the following and record your answers in scientific notation:
   a. 0.225 dm to mm
   b. 44,163 cs to ks
   c. 0.0000000000091 kL to microliter
   d. 2.5 mg to ng
   e. 20,190 microsecond to ms
   f. 7,000 nL to kL
7. How many significant figures are in each of the following numbers?
   a. 5.40 km
   b. 210 μL
   c. 801.5 g
   d. 1,000 ns
   e. 101.0100 mL
   f. 1.2×10^3 mm
   g. 0.00120 cg
   h. 0.0102 ds
   i. 9.010×10^{-6} ks
   j. 2,370.0 μg
8. Round off each of the following numbers to four significant figures:
   a. 15.9994 nL
   b. 0.66549 s
   c. 87,500 mg
   d. 1.0080 kL
   e. 48,859 μm
   f. 0.027225 s
9. Calculate the following, answer using proper significant figures and units:
   a. 43.00 dL + 19.1 dL
   b. (13.980 cg – 11.0 cg) + (340 cg + 217 cg)
   c. 2.7800×10^4 nm + 1.10×10^4 nm
   d. 0.00450 mm × 3.0454 mm × 1.0000 mm
   e. $\dfrac{0.0320m \times 9.100m \times 0.0045m}{4320m}$
   f. $\dfrac{(4230g + 547g)}{0.1340cm \times 0.050cm \times 3.10cm}$
   g. $(8.900 \times 10^8$ km$) \times (3.20 \times 10^{-5}$ km$)$

DENSITY

Answer the following questions. Remember to include units and significant figures in your answer.

1. Calculate the density of each of the following:
   a. 252 mL of a solution with a mass of 500 g

   b. A 6.75 g solid with a volume of 5.35 cm^3

   c. A substance with a mass of 7.55×10^4 kg and a volume of 9.50×10^3 L

2. Calculate the volume of each of the following:
   a. 26.5 g of a solution with a density of 7.48 g/mL

   b. A 3.400 kg solid with a density of 10.74 g/mL

3. Calculate the mass of each of the following:
   a. A solid with a volume of 1.68 L and a density of 9.2 g/mL

   b. An 80 mL aliquot of a solution with a density of 5.80 g/cm^3

   c. A solid cube with a density of 2.65 g/mL and dimensions of 0.50 m × 250 mm × 3.5 cm

SI BASE UNITS

1. What are some of the advantages to having a SI (metric) system of measurement?

2. Write out a unit of measurement that is not used in the SI (metric) system.

3. What is the SI base unit for length?

4. What is mass measured with?

5. Is volume a derived unit?

6. What is the SI base unit for volume?

7. Determine the SI base units for the following:
   a. time
   b. temperature
   c. length
   d. mass
   e. amount of substance

8. Describe each (how it is measured) and give a SI unit for each:
   a. mass
   b. length
   c. time
   d. temperature
   e. volume

SCIENTIFIC NOTATION

1. Change the following into scientific notation:
   a. 123,000,000,000,000
   b. 234,044,000
   c. 0.00000034
   d. 0.001002

2. Change the following into ordinary decimal notation:
   a. 5.432×10^{-5}
   b. 3.21×10^{1}
   c. 4.56×10^{12}
   d. 8.99×10^{-2}

3. Change the following into ordinary decimal notation:
   a. 5.67×10^{6}
   b. 3.4003×10^{-3}
   c. 8.8900×10^{-7}
   d. 7.71×10^{1}

4. Change the following into scientific notation:
   a. 0.0450000
   b. 3.21
   c. 45,800,000
   d. 230.01
   e. 0.00056700

5. Write 7.19×10^{-4} in decimal notation.

6. Write 8.34×10^{-3} in decimal notation.

METRIC CONVERSIONS

1. Match the following:
   i. 1 milligram
   ii. 1 centigram
   iii. 1 kilogram
   iv. 1 decigram
   v. 1 microgram
   vi. 1 nanogram

   a. 1×10^{-2} gram
   b. 1×10^{3} gram
   c. 1×10^{-3} gram
   d. 1×10^{-1} gram
   e. 1×10^{-6} gram
   f. 1×10^{-9} gram

2. Complete the following:
   a. 1 kilometer = meter
   b. 1 meter = decimeter
   c. 1 meter = centimeter
   d. 1 meter = millimeter
   e. 1 meter = micrometer
   f. 1 meter = nanometer

3. How many milligrams are in 1 gram?

4. How many decimeters are in 1 meter?

5. What is larger 1000 dg or 10 kg?

6. What is smaller 1 mL or 10 L?

7. Determine the following conversions (using dimensional analysis):
   a. How many kilograms are there in 50.0 grams?
   b. How many mL are there in 4.58 L?
   c. How many micrometers are there in 0.0000563 millimeters?
   d. How many deciseconds are there in 56,700 kiloseconds?

8. Determine the following conversions (using dimensional analysis):
   a. 6.7 g = kg
   b. 545.4 mm = cm
   c. 0.0458 ms = μs
   d. 5,987 m = dm
   e. 0.0303 mm = km
   f. 0.678 kg = cg
   g. 3.49 ng = μg

9. Convert the following and record your answers in scientific notation:
   a. 0.225 dm to mm
   b. 44,163 cs to ks
   c. 0.00000000000991 kL to μL
   d. 2.5 mg to ng
   e. 20,190 μs to ms
   f. 7,000 nL to kL

10. The unaided human eye has a resolving power of 0.1 mm. What is the equivalent resolving power in micrometers?

SIGNIFICANT FIGURES

1. A crucible is known to weigh 24.3162 g. Three students in the class determine the mass of the crucible by repeated massings on a simple balance. Using the following information, which student has done the most precise determination?

Student	Trial 1	Trial 2	Trial 3	Trial 4	Trial 5
A	24.8	24.9	24.8	24.9	24.8
B	24.8	24.0	24.2	24.1	24.3
C	24.5	24.1	24.5	24.1	24.3

2. Determine the number of significant figures in the following:
   a. 234.0012
   b. 0.004302
   c. 34.000
   d. 45.67900
   e. 300
   f. 3.200

3. How many significant figures are in 0.004050000?

4. When measuring the length of this line with the metric ruler provided, the first decimal place that is uncertain is:

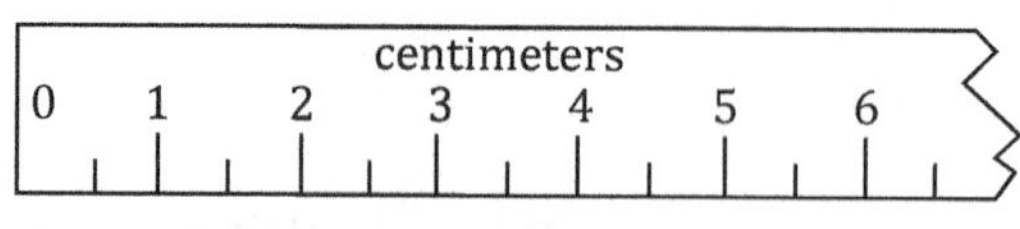

5. How many significant figures are in the measurement 102.400 meters?

6. The mass of a watch glass was measured four times. The masses were 99.997 g, 100.008 g, 100.011 g, and 100.005 g. What is the average mass of the watch glass?

7. Solve to the correct significant figures: 1.23 m × 0.89 m = ?

8. Round the following measurement to three significant figures: 0.90985 cm^2

9. Solve to the correct significant figures: 3.12 g + 0.8 g + 1.033 g = ?

10. When performing the calculation 34.530 g + 12.1 g + 1,222.34 g, how many significant figures will the final answer have?

11. Perform each calculation and express all answers in the correct number of significant figures:
    a. 9.40 *cm* × 2.6 *cm*
    b. $\frac{0.084g}{0.640mL}$
    c. $\frac{21.50g}{4.06cm \times 1.8cm \times 0.905cm}$
    d. $2.6 \times 10^5 + 4.1 \times 10^7$
    e. $\frac{\left(6.02 \times 10^2\right) \times 0.32000}{\left(5.325 + 21.0\right) \times \left(45.44 + 0.32\right)}$

12. How many significant figures are in each of the following numbers?
    a. 5.40 km
    b. 210 μL
    c. 801.5 g
    d. 1,000 ns
    e. 101.0100 mL
    f. 1.2×10^3 mm
    g. 0.00120 cg
    h. 0.0102 ds
    i. 9.010×10^{-6} ks
    j. 2,370.0 μg

13. Round off each of the following numbers to four significant figures:
    a. 15.9994 nL
    b. 0.66549 s
    c. 87,550 mg
    d. 1.0080 kL
    e. 4885 μm
    f. 0.027225 s

14. Perform the following mathematical operations and express your answers to the proper number of significant figures and units:
    a. 642 m × $(4.0 \times 10^{-5}$ m)
    b. 17 g ÷ $(3.88 \times 10^7$ cm^3)
    c. $(2.9 \times 10^{-5}$ cg) + $(8.1 \times 10^{-4}$ cg)
    d. $(4.3 \times 10^{-5}$ cm$)^3$
    e. $(5.40 \times 10^{-18}$ kg) / 769 mL
    f. [59 cm × $(3.24 \times 10^{-2}$ cm)] ÷ $(4.80 \times 10^4$ cm)
    g. [42 dm × $(4.02 \times 10^{23}$ dm)] ÷ 0.016 dm
    h. 123.040 kg – 55.06 kg
    i. 0.00000016 g / 74.3 L
    j. 10.0 ns + 54.600 ns

15. Using two different instruments, I measured the length of my foot to be 27 centimeters and 27.00 centimeters. Explain the difference between these two measurements.

16. Determine the number of significant figures in each of the following numbers:
    a. 23,450,000 μg
    b. 0.00340 kL
    c. 24.90000 dm
    d. 32,001,000 cs
    e. 3.4500×10^3 ng
    f. 0.102 mL

17. Calculate the following, answer using proper significant figures and units:
    a. 43.00 dL + 19.1 dL
    b. 6.500 ms + 120 ms
    c. (13.980 cg – 11.0 cg) + (340 cg + 217 cg)
    d. 2.7800×10^4 nm + 1.10×10^4 nm
    e. 0.00450 mm × 3.0454 mm × 1.0000 mm
    f. $\dfrac{0.0320m \times 9.100m \times 0.0045m}{4320m}$
    g. $\dfrac{0.00321g}{0.02000mL} \times 2390mL$
    h. $\dfrac{(4230g + 547g)}{0.1340cm \times 0.050cm \times 3.10cm}$
    i. $(8.900 \times 10^8$ km$) \times (3.20 \times 10^{-5}$ km$)$

TEMPERATURE

1. Convert 25.0°C to
    a. °F
    b. K

2. Convert 345 K to
    a. °F
    b. °C

3. Convert the following temperatures:
    a. 250 Kelvin to Celsius
    b. 17 Celsius to Kelvin
    c. 89.5 Fahrenheit to Celsius
    d. 339 Kelvin to Celsius
    e. 55 Celsius to Kelvin
    f. 383 Kelvin to Fahrenheit

4. Convert the following temperatures:
    a. 32°F to °C
    b. 988°C to °F
    c. -65.8°F to °C to K
    d. 243.8 K to °C to °F

5. What is the base unit for temperature?

6. What is the temperature in degree Celsius of 452 K?

7. What is the temperature in Kelvin of 67°F?

VOLUME AND DENSITY

1. Calculate the volume of a piece of metal that has a length of 4.5 cm, a height of 3.6 cm, and a width of 8.9 cm.

2. If a piece of copper has a volume of 51 cm^3 and a mass of 457.00 grams, what is its density in g/mL?

3. Calculate the volume (in mL) of a metal cube with the following dimensions: length = 4.5 cm, width = 6.7 cm, height = 2.4 cm

4. Calculate the volume (in cm^3) of a metal cylinder with the following dimensions: radius = 93.4 cm, height = 67800 cm

5. Calculate the volume of a metal cube (in cm^3) with the following dimensions: length = 0.98 dm, width = 990 mm, height = 3.3 cm

6. 56.7 mL of water was added to a 100.0 mL graduated cylinder. A solid sample was placed into the graduated cylinder, completely submerged in the water. The water level rose to 89.4 mL. What is the volume of the solid sample (in cm^3)?

7. Calculate the density (g/mL) of a solid sample having a mass of 367 grams and a volume of 89.4 cm^3.

8. Calculate the volume (in L) of a solid sample having a mass of 57.8 grams and a density of 1.26 g/cm^3.

9. The density of platinum metal is 21.5 g/cm^3. Calculate the mass (in g) of a block of platinum that has a volume of 0.876 L.

10. A lead block has a length of 0.22 meters, a width of 365 millimeters, a height of 4.8 decimeters, and a mass of 439 kilograms. Based on this information, what is the density of lead in g/cm^3?

11. If the density of alcohol is 0.789 g/mL, how many mL of alcohol must be measured out for an experiment that requires 2.50 grams of alcohol?

12. Suppose you wish to purchase a waterbed that has the dimensions 2.45 m × 21.5 dm × 23 cm. How many kilograms of water does this bed contain? (The density of water is 1.00 g/cm^3.)

13. The mass of a toy spoon is 7.5 grams, and its volume is 3.2 mL. What is the density of the toy spoon?

14. A mechanical pencil had the density of 3.0g/cm^3. When the pencil was placed into 56.8 mL of water, the water level rose to 73.1 mL (the pencil being completely submersed).
    a. What is the volume of the pencil?

    b. What is the mass of the pencil?

15. A screwdriver has the density of 5.5 g/mL. The mass of the screwdriver is 2.3 grams. What is the screwdriver's volume?

16. A metal cylinder has a mass of 5.48 grams. The measured height is 5.6 cm and diameter is 445 dm.
    a. What is the volume of the metal cylinder?

    b. What is the density of the metal cylinder?

17. What is the volume of a metal cube with the following dimensions?
    length = 3.2 cm, width = 2.1 cm, height = 1.5 cm
    volume = length × width × height

18. What is the definition of density?

19. What is the density of a metal if a 15.4 gram sample has a volume of 1.96 cm^3?

20. 5.00 mL = ______________ cm^3?

21. 2.5 dL = ______________ cm^3?

22. What is the volume of a rectangular solid having the following dimensions?
length = 8.30 cm, width = 3450 mm, height = 0.0540 m

23. What volume of a liquid having a density of 3.48 g/cm^3 is needed to supply 5.00 grams of the liquid?

24. A metal sample having a mass of 30.9 grams was added to a graduated cylinder containing 23.2 mL of water. The volume of the water plus the sample was 24.8 mL. What is the density of the metal?

25. An unknown liquid has a density of 0.786 liters and a density of 19.4 g/cm^3. What is the mass of the liquid?

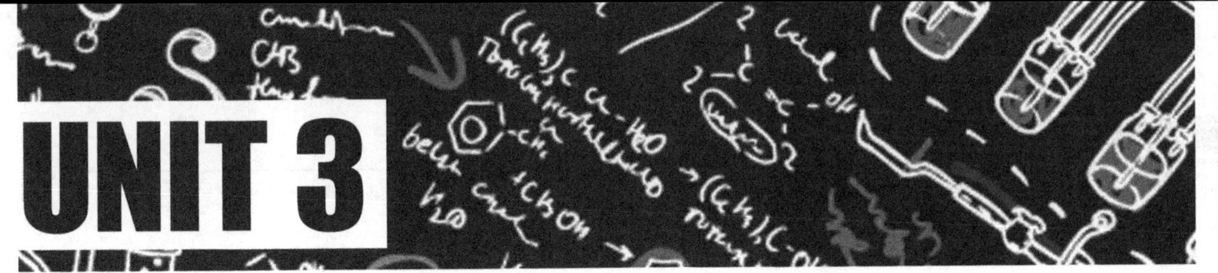

THE AMAZING ATOM (THE ATOM AND THE PERIODIC TABLE)

Atom's Timeline
- The Atom
- The Greek's Atom
- Dalton Brings Back the Atom
  - Dalton's Atomic Theory

Subatomic Particles
- The Discovery of the Electron
- The Discovery of the Proton
- The Discovery of the Neutron

Structure of the Atom—Its Discovery
- First Theory: J. J. Thomson's Plum Pudding Model
- Second Theory: Ernest Rutherford's Gold Foil Experiment

Structure of the Atom—Protons, Neutrons, Electrons
- When Atoms Become Ions
- Isotopes—They're Just Atoms!
- Atomic Mass

The Periodic Table and Elemental Symbols
- Early Classification
- The Modern Periodic Table
  - Horizontal Rows are Called Periods and Vertical Columns are Called Groups or Families
  - Group 1 (i)—Alkali Metals
  - Group 2 (ii)—Alkaline Earth Metals
  - Group 17 (vii)—Halogens
  - Group 18 (viii)—Noble Gases
  - Period 6—Lanthanide Series
  - Period 7—Actinide Series

OBJECTIVES FOR UNIT 3

- To be familiar with Dalton's atomic theory.
- To understand the history of the atom and its discoverers.
- To know and understand the important discoveries of the subatomic particles.
- To understand the concept of isotopes and to be able to calculate relative atomic masses from atomic masses and isotopic distribution.
- To write or interpret chemical symbols for isotopes.
- To understand the concept of isotopes and to be able to calculate relative atomic masses.
- To be familiar with aspects of the periodic table, including metals and nonmetals, periods, and groups.

ATOM'S TIMELINE

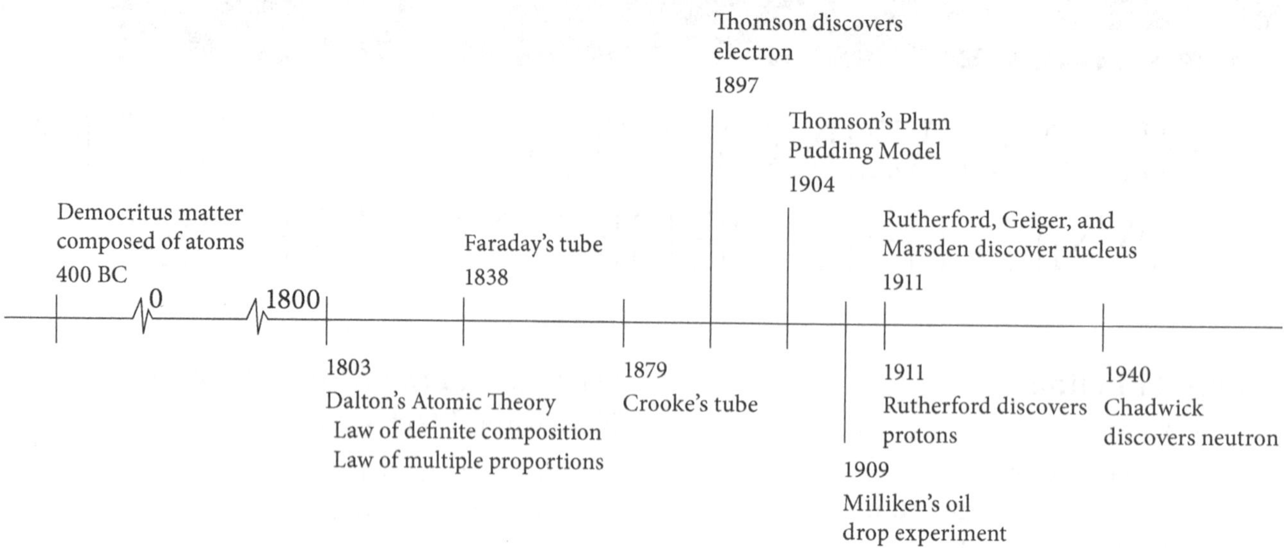

THE ATOM

All elements are composed of an extremely small particle called the atom. This fact has been proven by our current technologies.

How did we derive our understanding the existence of the atom?

© Georgios Kollidas/Shutterstock.com

THE GREEK'S ATOM

Let's take a little ride back into history to Greece in 500 BC. Here we meet a teacher named Leucippus and his student named **Democritus**. These two men were philosophers, speculating on the nature of matter. Their observations in nature led them to believe that matter consists of tiny particles of various kinds, separated by space through which these particles traveled.[1] Democritus gave the name atoms to these particles. A few years later, a famous Greek philosopher named Socrates denounced the idea of atoms, finding more interested in abstract ideas. Socrates's ideas carried to his student Plato and to Plato's student Aristotle. So the concept of atoms was put to rest for almost 2,000 years.

DALTON BRINGS BACK THE ATOM

Now move your time clock to the beginning of the 1800s. The works of Priestley, Scheele, Cavendish, and Lavoisier discovered oxygen and disproved the phlogiston theory. There began a movement away from Aristotle's four element theory and new elements and compounds started to be discovered. Air was found to contain four gases: oxygen, nitrogen, carbon dioxide, and water vapor.

© Georgios Kollidas/Shutterstock.com

Then along came a young Quaker from England named John Dalton. While studying the atmosphere, Dalton found all his air samples contained these four gases of oxygen, nitrogen, carbon dioxide, and water vapor. This made him wonder why the heavier carbon dioxide gas didn't settle down to the bottom of the atmosphere. He wondered if all the gases in air are mixed as one big compound or one big solution. Dalton made diagrams and read books on recent and past discoveries and ideas. It was when he read about Leucippus and Democritus that he began to wonder about the truth behind atoms. After much deep thought, the atomic theory was revealed to him. Interestingly, he didn't use experimentation to formulate his theory.

DALTON'S ATOMIC THEORY

John Dalton developed the atomic theory in 1803, most of which is accepted today. It is stated as follows:

© HuHu/Shutterstock.com

1. Each element is made up of tiny, individual particles called atoms.
2. Atoms are indivisible; they cannot be created or destroyed.

   Our current adaptation of this postulate is the law of conservation of mass, which states: Atoms can neither be created nor destroyed in a chemical reaction, and as a consequence the total mass remains unchanged.
3. All atoms of each element are identical in every aspect.

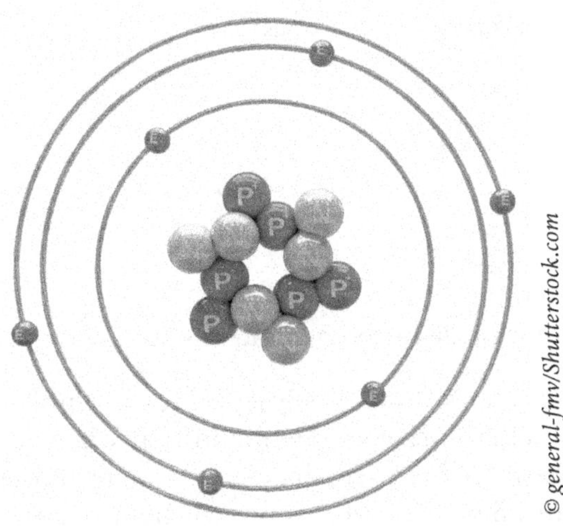

© general-fmv/Shutterstock.com

4. Atoms of one element are different from atoms of any other element.

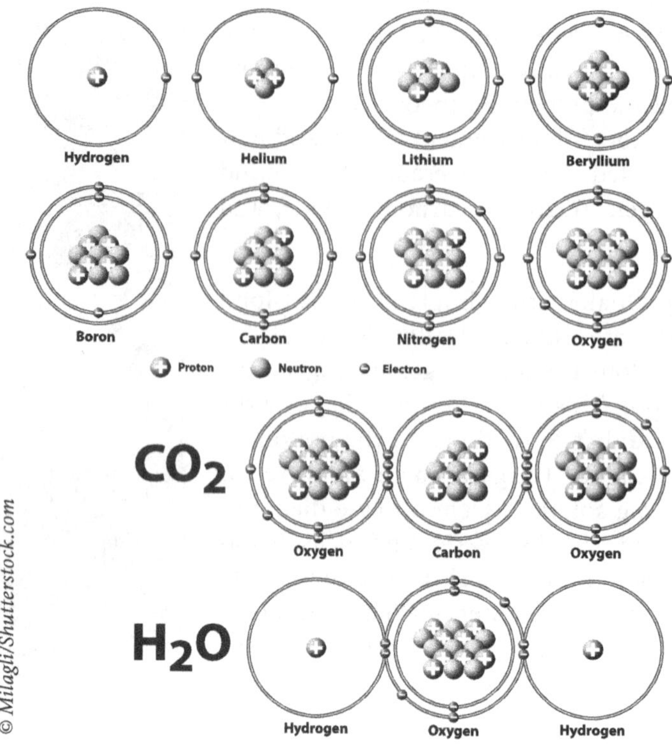

5. Atoms of one element may combine with atoms of another element, usually in the ratio of small whole numbers, to form chemical compounds.

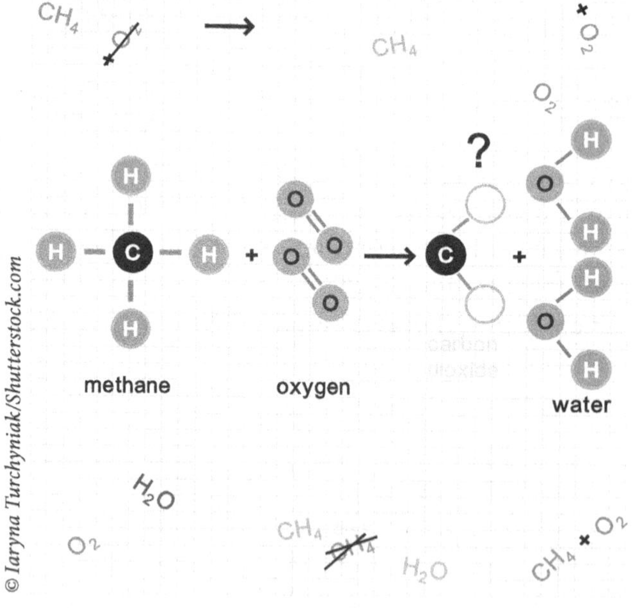

Dalton had spherical models of atoms made to demonstrate his atomic theory.

One man who was intrigued with Dalton's atoms was Joseph Proust. Unlike Dalton, Proust relied on laboratory observations for explaining matter's behavior. He found that when elements combine to form chemical compounds, the elements unite in definite proportions by weight. Proust believed this observation is a result of elements containing atoms. So when water is formed by hydrogen and oxygen, it will always take

11% hydrogen and 89% oxygen. We call this the law of definite composition. This law of definite composition remains the fundamental principle of the science of chemistry.

Dalton also found it necessary to have relative weights given to these atoms. At the time, hydrogen was discovered as being the lightest element; therefore Dalton gave the atoms of hydrogen the relative mass of 1. All other atoms were compared to hydrogen atoms and given a relative mass. Because Dalton believed water to be formed from one atom hydrogen and one atom oxygen (is this still true?), he gave oxygen atoms the relative mass of 7.

Out of this work Dalton discovered that the compounds that we know as carbon monoxide and carbon dioxide have the following ratio: carbon monoxide: 1 atom carbon combines with 1 atom oxygen.

Carbon dioxide: 1 atom carbon combines with 2 atoms oxygen

This discovery led to another fundamental law in chemistry: the law of multiple proportions that states that when two elements are combined to form more than one compound, the different masses of one element that combine with the same mass of the other element are in a simple ratio of whole numbers.

Example:

Carbon monoxide and carbon dioxide are both composed of the carbon and oxygen atoms, but in different proportions.

	Carbon Monoxide (CO)	Carbon Dioxide (CO_2)
The Elements	3.0 g carbon (C) + 4.0 g oxygen (O)	3.0 g carbon (C) + 8.0 g oxygen (O)
The Compound	7.0 g carbon monoxide (CO)	11.0 g carbon dioxide (CO_2)
Mass Ratio	4.0 g O : 3.0 g C	8.0 g O : 3.0 g C
Comparing	4 g O in CO : 8 g O in CO_2, thus it is a 1 : 2 ratio	

SUBATOMIC PARTICLES

THE DISCOVERY OF THE ELECTRON

In 1838, Michael Faraday, an Englishman with little formal schooling, made one of the most important contributions into the fields of electricity and magnetism. He was the first to report induction of an electric current from a magnetic field. Faraday invented the first electric motor and dynamo, demonstrated the relation between electricity and chemical bonding, discovered the effect of magnetism on light, and discovered and named diamagnetism. He also provided the experimental, and much of the theoretical, foundation on which J. C. Maxwell built his electromagnetic field theory.

In one fundamental experiment, Faraday was the first to see proof to the existence of the electron (although he didn't realize it at the time). He took a tube with most of the air removed and discovered a purple glow appeared inside when hooked to an electric current. Upon better perfection of this tube (better removal of air), he found the tube carried a fluorescent light inside when an electric current was applied.

© Ensuper/Shutterstock.com

© Jose Angel Astor Rocha/Shutterstock.com

William Crookes and other English scientists in 1879 worked on perfecting Faraday's experiment with an "airless tube." Crookes removed even more air than Faraday had, leaving only a few molecules remaining. Instead of a purple glow, Crookes created a fluorescent light when he used an induction coil to discharge a high voltage current of electricity through the tube. An English apothecary and physician named William Watson performed a similar experiment a century and a half before. Crookes also saw that the fluorescent light (cathode rays) would bend when a strong magnetic field was applied. This made Crookes wonder if this was just light or if it were matter? Crookes decided that these cathode rays are a fourth state of matter, not gas, liquid, nor solid but "radiant matter." Because of this, Crookes missed discovering the electron.

Faraday's and Crooke's work led to the discovery of cathode rays. Today we know these rays as a stream of electrons leaving the negative electrode, or cathode, in a discharge tube (an electron tube that contains gas or vapor at low pressure), or emitted by a heated filament in certain electron tubes. Cathode rays cause fluorescent materials to luminance and are utilized in:

oscilloscopes

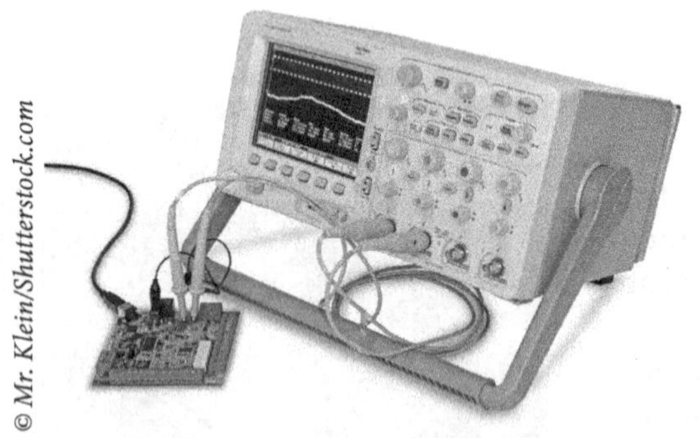

© Mr. Klein/Shutterstock.com

television tubes

© HomeArt/Shutterstock.com
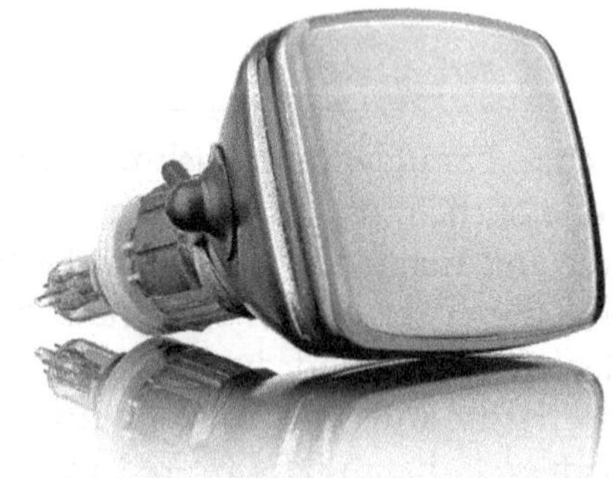

By the 1890s, scientists had compiled the following observations about cathode rays:

1. The rays are emitted from the cathode when electricity is passed through an evacuated tube.
2. The rays are emitted in a direction perpendicular to the cathode surface.
3. Cathode rays travel in straight lines.
4. Cathode rays cause glass and other materials to give off a fluorescent glow.
5. Cathode rays are deflected in a magnetic field in the way expected for negatively charged particles.
6. The properties of cathode rays do not depend on the composition of the cathode. For example, the cathode rays from an aluminum cathode are the same as those from a silver cathode.

J. J. Thomson, a young English scientist who was the head of Cavendish Laboratories at Cambridge, reproduced Crookes experiment, varying different conditions such as removing as much gas as possible, changing the electric charge, and changing the magnetic field applied. Still he was pondering this mysterious behavior of the cathode rays. Finally, in 1897, Thomson declared that cathode rays "are particles of negative electricity."[2] He determined that the stream of cathode rays, which the magnet had deflected, was made up of electrons (corpuscles as Thomson first called them), torn away from the atom of the gas in the tube.

- Thomson proposed that every atom is composed of these electrons
- These electrons move at a speed of 160,000 miles per second.
- The electron was the smallest particle of matter.
- Knowing the amount of magnetic charge applied and how it affected the bending of the cathode rays, Thomson was able to determine a ratio of the electric charge of the electron to the mass—the "e/m." Through amazing experimentation, Thomson was able to determine the mass of the electron to be two thousand times less than that of the atom hydrogen, the lightest substance known.

© catwalker/Shutterstock.com

In 1909 at the University of Chicago, **Robert Millikan** isolated an electron and was able to conclude the negative charge of the electron. His famous experiment is known as the oil drop experiment. This experiment used an apparatus consisting of a top and bottom brass plate about one-third of an inch apart with transparent sides. He made a hole the diameter of a needle in the top brass plate. A strong beam of light passed between the plates. He connected the two brass plates to a ten thousand volt battery. Millikan then used a commercial atomizer to spray ten-thousandth of an inch of oil droplets into the air above the upper plate. After hours of waiting, a single drop of oil finally fell through the hole on the upper plate (Millikan saw this while looking through a microscope). After this, he used a sample of radium to remove electrons from the neutral oil drop. Before the oil drops were treated with the radium, they would just fall down from the top brass plate to the bottom brass plate, regardless of any change to the battery set up or charge. After the treatment with radium, Millikan observed the oil drops instantly began to slow down in their fall between the plates. Millikan knew, "the droplet was no longer neutral; it had lost some of its electrons and become positively charged."[3] Millikan could also determine the amount of electrons lost based upon the speed of the "charged" oil droplet.

Millikan, in his famous oil drop experiment, agreed with J. J. Thomson's mass value on the electron and provided the value of *e*, the electron charge to be: $e = 1.602 \times 10^{-19}$ coulomb (C). For his study of the electronic charge and the photoelectric effect, Millikan won the 1923 Nobel Prize for Physics.

In 1911, at the request of Thomson, a fellow at Cavendish laboratories named Charles Thomson Rees Wilson (known as C.T.R.), created a camera that took a picture of a single electron surrounded by a droplet of water.

© naum/Shutterstock.com

THE DISCOVERY OF THE PROTON

At age 24, **Ernest Rutherford** from New Zealand went to work at Cavendish Laboratories in 1895. Four years later, J. J. Thomson assigned Rutherford as chair of physics at McGill University in Montreal.

In 1902 Rutherford, with another young scientist at McGill named Frederick Soddy, published that radioactivity involves atoms losing positive particles that they called alpha particles.

Rutherford discovered that alpha particles were actually positive atoms of helium. He found this out by sealing radium in a tube and placing it in another tube, with no other gases (including air) present. After two days he examined the outer tube and found that it contained positive helium atoms. These alpha particles Rutherford identified as positive atoms of helium, ejected from radium with a velocity of 12,000 miles per second, a speed which would bring us to the sun, 93 million miles away, in a little more than two hours.[4]

In 1911, Rutherford studied these alpha particles (positive helium atoms) by passing them through nitrogen gas. Rutherford found two amazing discoveries. First, the alpha particles were hitting something large in the nitrogen, not allowing the entire alpha particle to pass in a straight-line path. This gave evidence of a possible nucleus in the nitrogen atoms (discussed later). Second, Rutherford found a group of positively charged hydrogen atoms after the particles were sent through the nitrogen gas. Twenty-two years after Thomson discovered the electron, Rutherford announced his discovery of the proton (named from the Greek word *protos-* meaning first). In 1919, Rutherford went to Cambridge to become director of the Cavendish Laboratory.

Interestingly, Rutherford actually performed the first artificial transmutation in history. When Rutherford bombarded nitrogen with alpha particles (helium atoms) and positive hydrogen (protons) were released, nitrogen was changed into oxygen.

Nitrogen + Helium (positively charged) → Hydrogen (positively charged) + Oxygen

Radioactivity Penetration Range

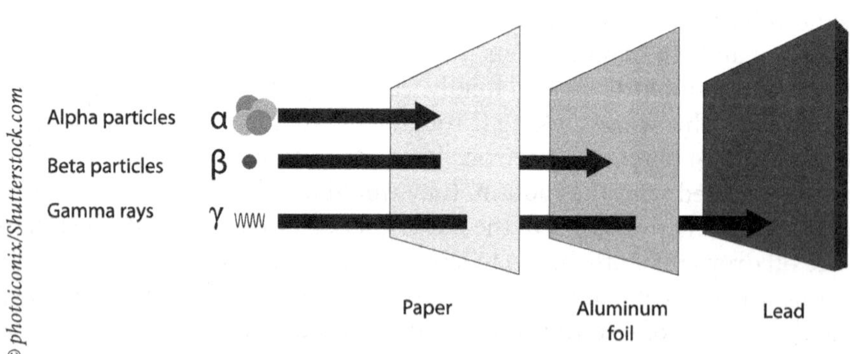

© photoiconix/Shutterstock.com

THE DISCOVERY OF THE NEUTRON

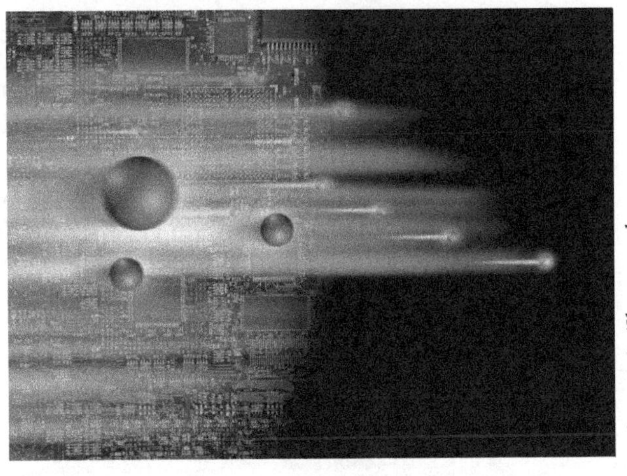
© Anteromite/Shutterstock.com

In 1932, an Englishman named **James Chadwick**, an assistant to Rutherford at Cavendish Laboratory, found the existence of the neutron. For its discovery, Chadwick put a piece of beryllium in a vacuum chamber with some polonium. The polonium emitted alpha rays, which struck the beryllium. When struck, the beryllium emitted neutral rays. Chadwick put a target in the path of these rays. When the rays hit the target, they knocked atoms out of it. The atoms, which became electrically charged in the collision, flew into a detector. Chadwick's used the detector to count these atoms and estimate their speed. Chadwick used targets of different elements, measuring the energy needed to eject the atoms of each. Chadwick concluded that the speed of these atoms was a result of neutral particles (neutrons). To prove that the particle was indeed the neutron, Chadwick measured its mass. Since he could not weigh it directly, he measured everything else in the collision and used that information to calculate the mass.

STRUCTURE OF THE ATOM—ITS DISCOVERY

FIRST THEORY: J. J. THOMSON'S PLUM PUDDING MODEL

In 1904, after his discovery of the electron, Thomson made an attempt to explain how the electrons were situated in the atom. At this time, neither the actual mass nor the charge of the electron was known. Since atoms were electrically neutral and were known to contain electrons, they must also contain positive-charged material.

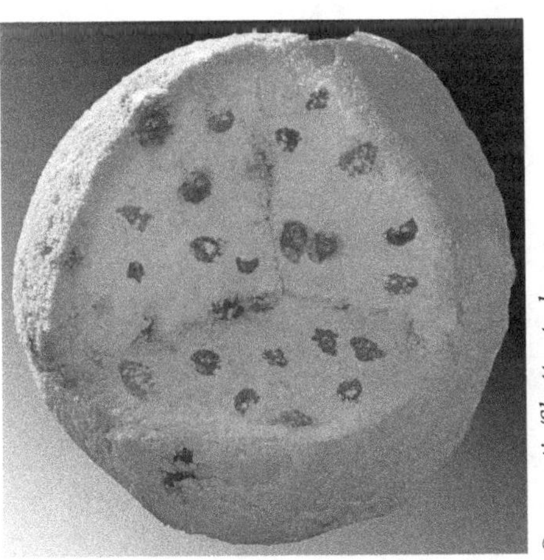
© magnetix/Shutterstock.com

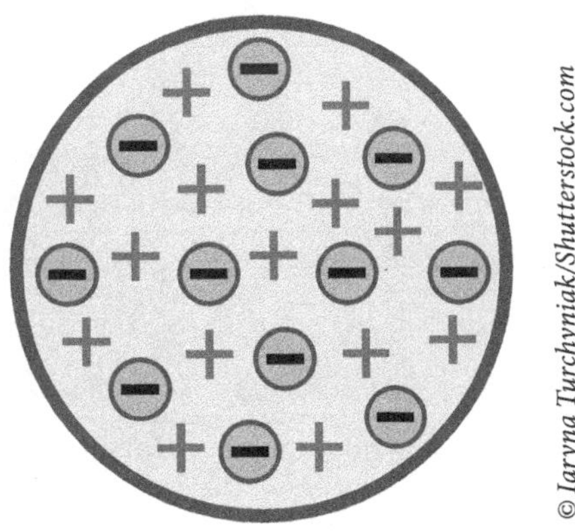
© Iaryna Turchyniak/Shutterstock.com

From this information, Thomson proposed the "plum pudding" model of the atom. In this model, the atom was considered to be a sphere of positive charge in which an equivalent negative charge was scattered. The electrons were believed to be spread through the positive sphere like raisins in a plum pudding. Thomson was unable to suggest how the electrons were held in the positive sphere. His plum pudding model could explain the neutrality of an atom, the origin of electrons and chemical properties, but could not account for spectral line or radioactivity. However his suggestion affected Bohr's atomic theory.

SECOND THEORY: ERNEST RUTHERFORD'S GOLD FOIL EXPERIMENT

In 1911, shortly after Rutherford discovered the proton by shooting alpha particles (charged helium atoms) through nitrogen gas, he recruited his assistance Hans Geiger (developer of the Geiger counter—a detector and counter of fast particles) and E. Marsden to continue the investigation. They bombarded other materials with alpha particles. One of the materials chosen was gold foil, which led to the famous gold foil experiment, a Rutherford experiment that gave final evidence to the existence of a nucleus in an atom.

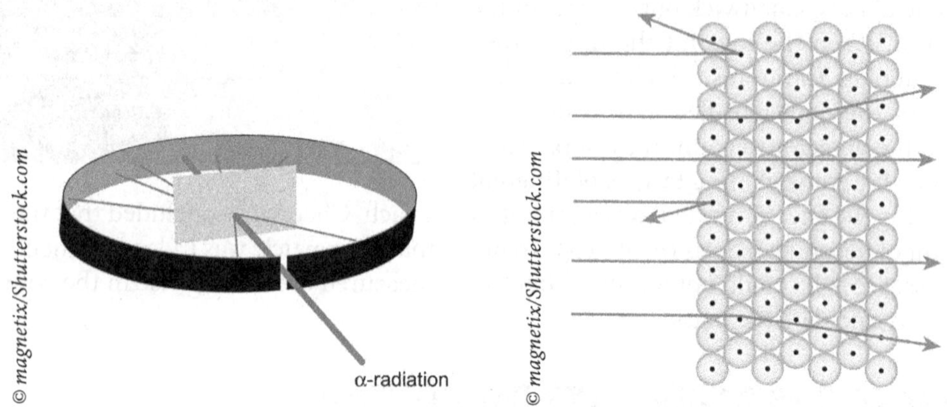

© magnetix/Shutterstock.com

© magnetix/Shutterstock.com

Much like the nitrogen gas experiment, what the scientists found in the gold foil experiment was some of the alpha particles were deflected, some at very large angles. Some were even deflected backwards. As Rutherford remarked, "It was as incredible as if you fired a 15-inch shell at a piece of tissue paper and it came back and hit you."

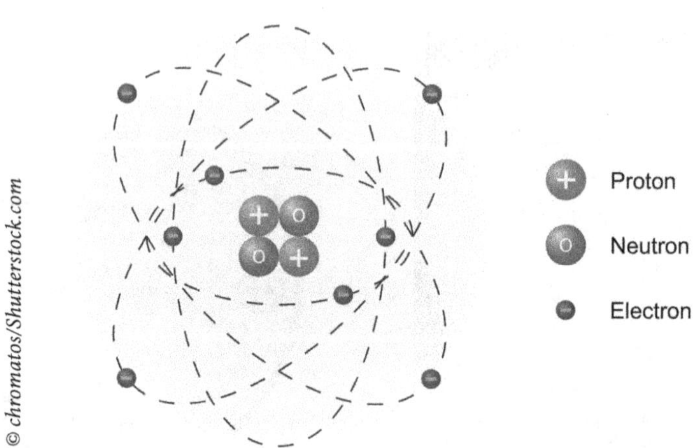

© chromatos/Shutterstock.com

STRUCTURE OF THE ATOM—PROTONS, NEUTRONS, ELECTRONS

The atom is composed of a nucleus containing protons and neutrons. The electrons exist outside the nucleus.

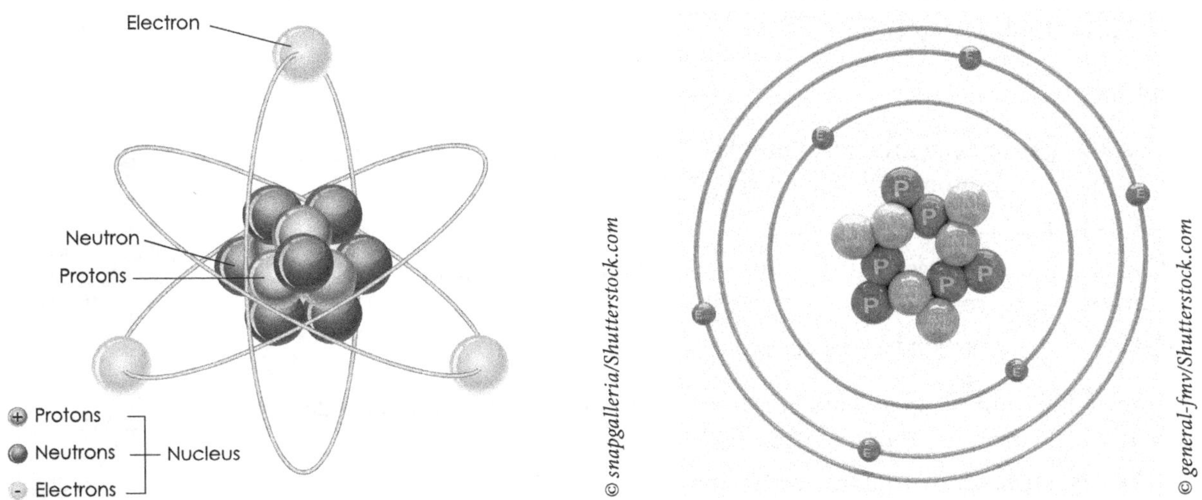

Other definitions associated with subatomic particles (protons, neutrons, electrons) include:

1. The atomic number is equal to the number of protons in the nucleus. Z ($_{Z}E$)
   - This number never changes for a particular atom (element).
   - You can find the atomic number on the periodic table.
   - When an atom is electronically neutral $\#p^+ = \#e^-$.

Examples:

- Chlorine: $_{17}\textbf{Cl}$ indicates the atomic number is 17 and chlorine has 17 protons.
  $_{17}\textbf{Cl}$ is neutral, chlorine has 17 electrons.
- Iron: $_{26}\textbf{Fe}$ indicates the atomic number is 26 and iron has 26 protons.
  $_{26}\textbf{Fe}$ is neutral, iron has 26 electrons.
- Copper: $_{29}\textbf{Cu}^{2+}$ indicates the atomic number is 29 and copper has 29 protons.
  $_{29}\textbf{Cu}^{2+}$ is not neutral, copper has 27 electrons.

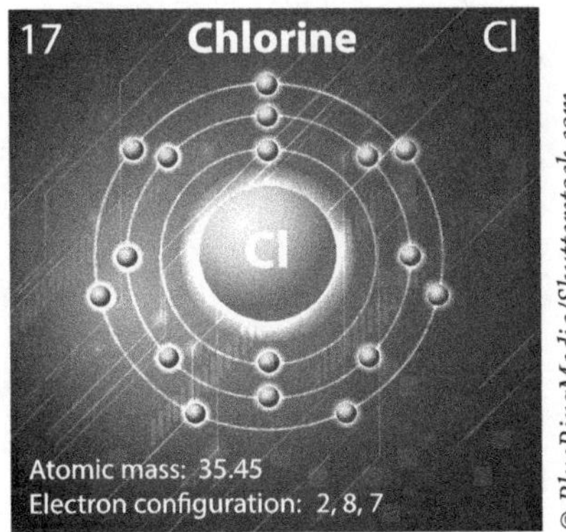

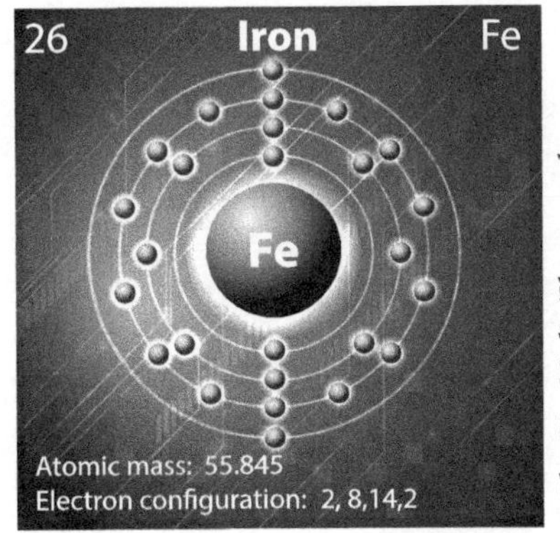

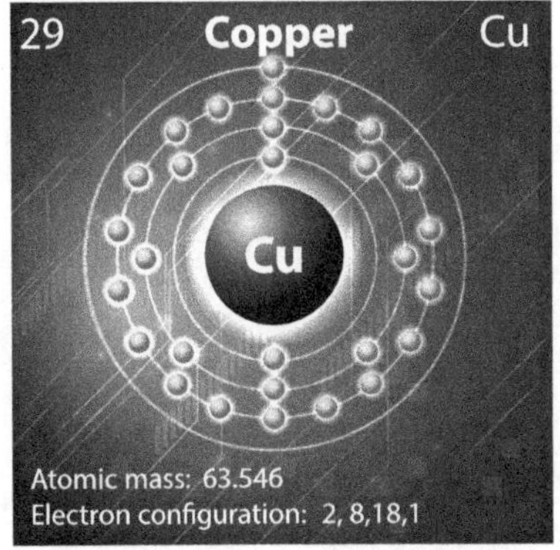

2. The mass number is equal to the sum of the number of protons and the number of neutrons in the nucleus: a (^{A}E)
   - This number never can change for a particular atom (element).
   - You cannot find the mass number on the periodic table, but there may be a similar number.
   - Mass number = $\#p^{+} + \#n$

Examples:

- Carbon: $^{13}_{6}\mathbf{C}$ indicates the mass number is 13 and the atomic number is 6. Carbon has 6 protons and 7 neutrons (13–6).

  $^{13}_{6}\mathbf{C}$ is neutral; carbon has 6 electrons.
- Carbon: $^{12}_{6}\mathbf{C}$ indicates the mass number is 12 and the atomic number is 6. Carbon has 6 protons and 6 neutrons (12–6).

  $^{12}_{6}\mathbf{C}$ is neutral; carbon has 6 electrons.

© Nerthuz/Shutterstock.com

- Iron: $^{57}_{26}\mathbf{Fe}^{2+}$ indicates the mass number is 57 and the atomic number is 26. Iron has 26 protons and 31 neutrons (57–26).

  $^{57}_{26}\mathbf{Fe}^{2+}$ is not neutral; iron has 26 electrons.

© Nerthuz/Shutterstock.com

WHEN ATOMS BECOME IONS

When an atom gains or losses an electron, it becomes an ion: charge (E^{charge})

1. Cation is a positively charged ion.

   Examples:

   Hydrogen ion H^{+}, carries a charge of +1 (is actually a proton most hydrogen has no neutron), has lost 1 electron

   Alpha Particle: A helium ion He^{2+}, carries a charge of +2, has lost 2 electrons.

   Others: Na^{+}, Li^{+}, Ca^{2+}, Cr^{2+}

2. Anion is a negatively charged ion.

   Examples:

   Chlorine ion Cl^{-}, which carries a charge of –1, has gained 1 electron.

   Sulfur ion S^{2-}, carries a charge of –2, has gained 2 electrons.

   Others: F^{-}, O^{2-}, N^{3-}

ISOTOPES—THEY'RE JUST ATOMS!

Going all the way back to the discovery of oxygen, leading up to the discovery of the atom, the "weights" of elements was being determined. As measurement and instrumentation improved, data became more accurate. Even with precision measurements, the recorded weights of elements varied from scientist to scientist, even from sample to sample. In 1912, Theodore W. Richards, the first American to win the Noble Prize in Chemistry, even found ordinary lead atoms to have to have a relative weight of 207.2, while lead atoms taken from a radioactive uranium ore from Norway was 206.05.[5] Even more confusing was the fact that these relative weights were not whole numbers (even though the neutron was not yet discovered). It was Fredrick Soddy, the same person who worked with Rutherford, who coined the term isotopes (meaning equal places) for the same elements having different weights. We now know that Dalton's statement that "all atoms of the same element have the same mass" is not completely true. Atoms can vary in the amount of neutrons they have. This will result in differing masses (atomic mass = $\#p^+ + \#n^o$) Therefore, these atoms we call isotopes.

Isotopes are atoms having different atomic masses or mass numbers, but the same atomic numbers.

Examples:

- Isotopes of chlorine:

  $^{37}_{17}\mathbf{Cl}$ indicates the atomic number is 17 and the mass number is 35.

  $^{37}_{17}\mathbf{Cl}$ indicates the atomic number is 17 and the mass number is 37.

  Knowing the amount of each isotope in a sample of chlorine, you will be able to determine the atomic mass for chlorine to be 35.45.

ATOMIC MASS

When you look on the periodic table, you notice a number called the atomic mass. The atomic mass is similar to the mass number. These two numbers are not the same. The atomic mass of an element is the average of the masses of the naturally occurring isotopes of that element (based on the amount of each isotope for a particular element).

The unit used to write atomic masses is amu (average atomic mass units).

To calculate the atomic mass, two items are needed:

1. The atomic masses of the isotopes of the element
2. The naturally occurring fractional abundances of the isotopes.

Example:

Isotope	Symbol	Percent Abundance	Mass (amu)
Carbon-12	^{12}C	98.89%	12.000000
Carbon-13	^{13}C	1.11%	13.003354

average atomic mass: 12.01 amu

Calculating average atomic mass examples:

Nickel is used primarily in the manufacture of metal alloys, such steel and nickel-chrome resistance wires. The three most abundant isotopes of elemental nickel are ^{58}Ni, ^{60}Ni, and ^{62}Ni.

28
Ni
Nickel
58.6934

Isotope	Percent Abundance	Mass (amu)
^{58}Ni	68.27%	58.00
^{60}Ni	26.10%	60.00
^{62}Ni	3.59%	62.00

To calculate the average atomic mass of nickel:

$$(0.6827 \times 58.00) + (0.2610 \times 60.00) + (0.0359 \times 62.00) = 57.48 \text{ amu}$$

THE PERIODIC TABLE AND ELEMENTAL SYMBOLS

EARLY CLASSIFICATION

- Late 18th Century: Lavoisier's list of 28 elements.
- Early 19th Century: John Dalton's list of approximately 60 elements.
- Dobereiner grouped three elements together with related properties.
- John Newlands grouped elements in groups of eight.
- J. Lothar Meyer arranged elements with similar physical properties.
- In 1869, **Dmitri Mendeleev**, a Russian scientist, arranged elements in order of increasing atomic mass. He also placed the elements with similar properties in "groups." He is considered the father of the periodic table.

© Olga Popova/Shutterstock.com

In 1912, Henry G. Moseley discovered the law of atomic numbers. He created a table of elements arranged in order of atomic numbers. At the time there were 92 elements (1st—hydrogen, 92nd—uranium).

THE MODERN PERIODIC TABLE

The last major change to the periodic table resulted from Glenn Seaborg's work in the middle of the 20th century. Starting with plutonium in 1940, Seaborg discovered actinides 94 to 102 and reconfigured the periodic table by placing the lanthanide/actinide series at the bottom of the table. In 1951 Seaborg was awarded the Nobel Prize in chemistry and element 106 was later named seaborgium (Sg) in his honor.

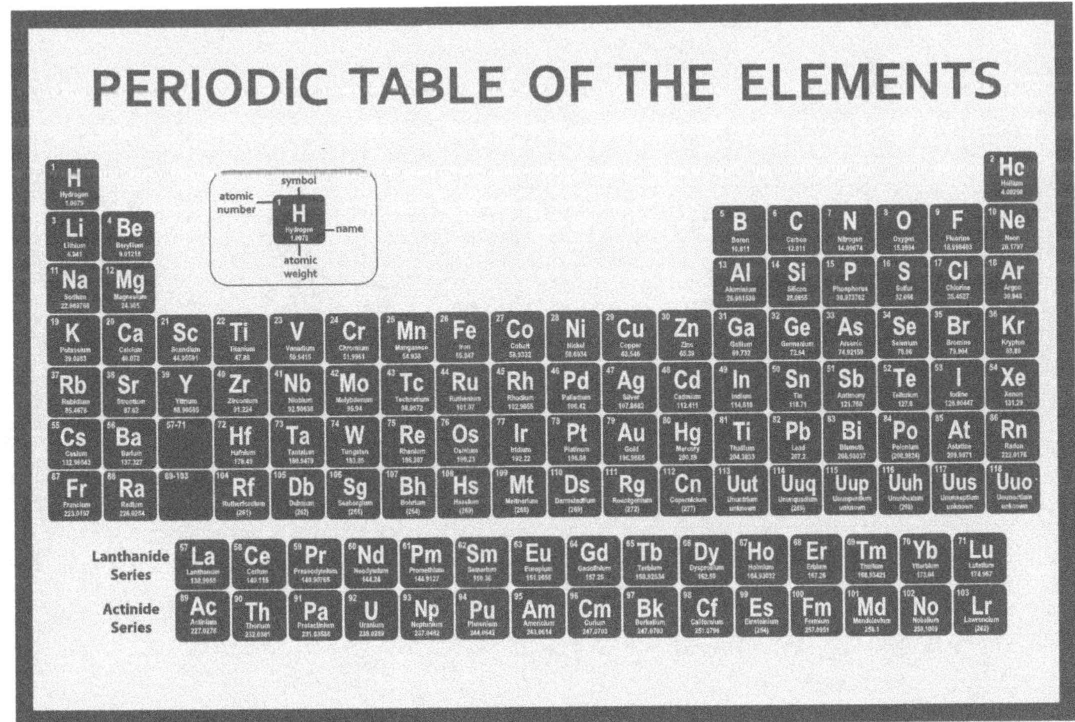

© okili77/Shutterstock.com

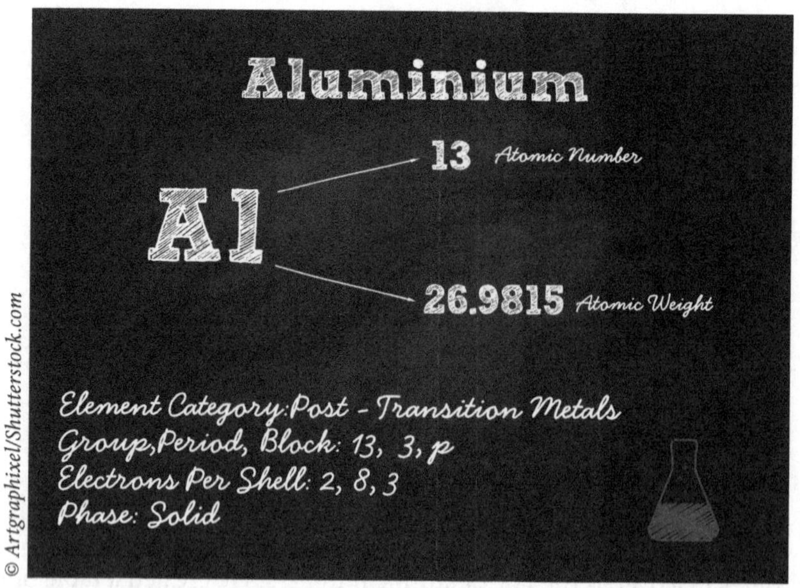

As of April 2015, 118 elements are currently known.

- 87 are metals
- 38 are radioactive
- 20 are man-made (all radioactive)
- 11 occur as gases
- 2 occur as liquids

HORIZONTAL ROWS ARE CALLED PERIODS AND VERTICAL COLUMNS ARE CALLED GROUPS OR FAMILIES

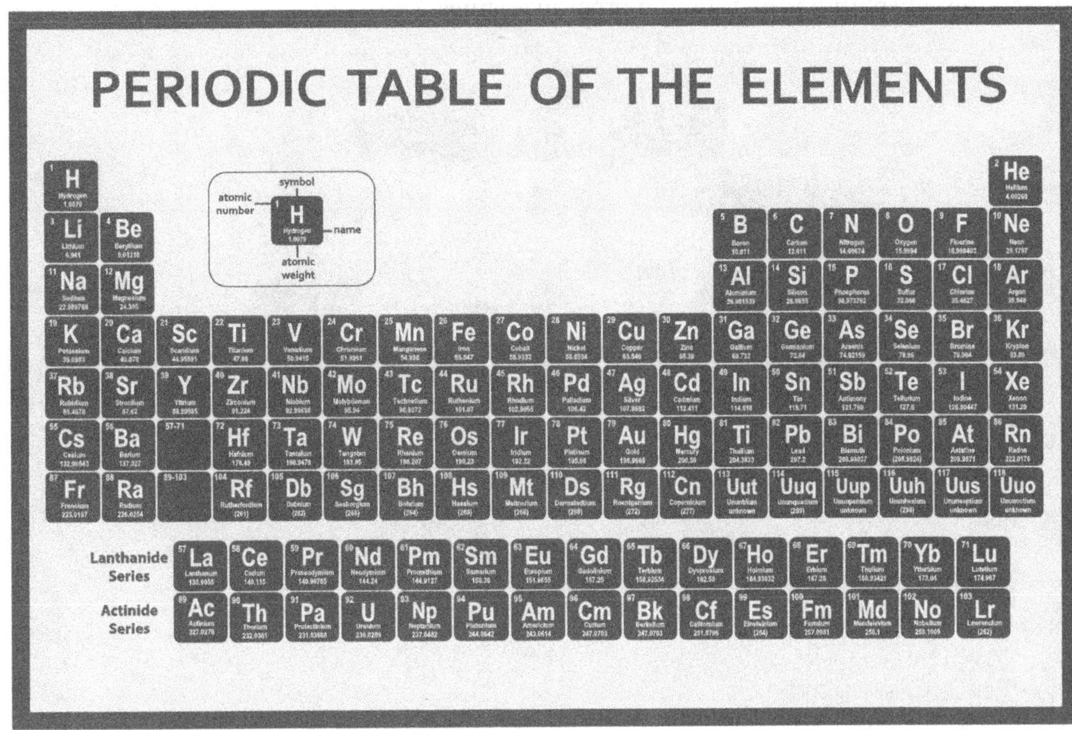

© okili77/Shutterstock.com

GROUP 1 (I)—ALKALI METALS

Example: Potassium, used in fertilizers, fireworks, and liquid detergents.

© Africa Studio/Shutterstock.com

GROUP 2 (II)—ALKALINE EARTH METALS

Example: Magnesium is used in pyrotechnics, alloys for aircrafts, car engines, and missile construction. Magnesium is also an ingredient in Epsom salts and other medicines.

GROUP 17 (VII)—HALOGENS

Example: Bromine is used in petrol antiknock compounds, flame proofing agents, water purification compounds, dyes, medicines, photography, and pesticides.

GROUP 18 (VIII)—NOBLE GASES

Example: Krypton is used with argon in fluorescent lights and is used in photographic flash lamps.

Periods at the bottom of the periodic table:

PERIOD 6—LANTHANIDE SERIES

Example: Lanthanum is used in the carbon lighting applications for studio lighting and projection, used in making special optical glasses, lighter flints, and alloys.

© Zerbor/Shutterstock.com

PERIOD 7—ACTINIDE SERIES

Example: Uranium is used in nuclear fuel, nuclear reactors, and high energy X-rays.

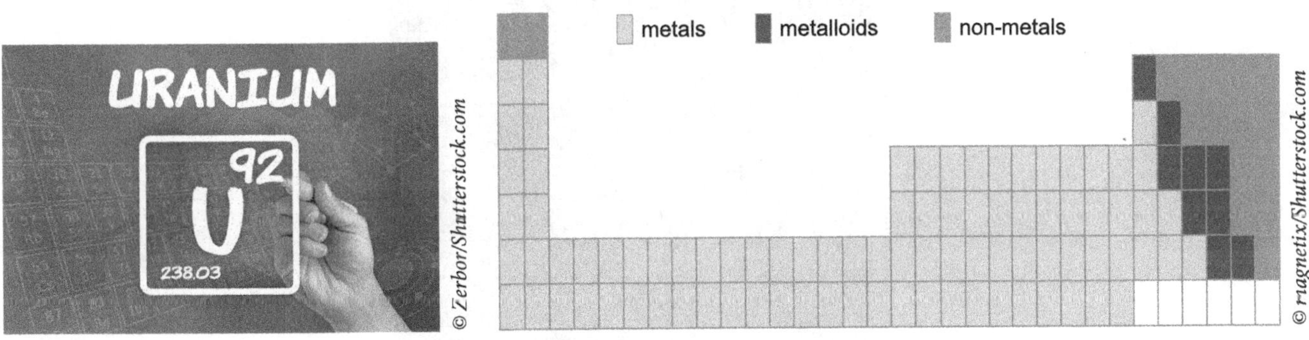

© Zerbor/Shutterstock.com

© magnetix/Shutterstock.com

Metals have a high luster, are good conductors of electricity, malleable, ductile, have high densities, high melting points, and hardness.

Free state metals: Au, Ag, Cu, Pt

© Oleksiy Mark/Shutterstock.com

Nonmetals are dull, poor conductors of electricity, brittle, have low densities, low melting points, are soft, and combine with metals and with each other.

Examples: Solids: S, P, C Liquids: Br Gases: H, F, Cl, Ne

Metalloids are located on the stairstep line, except aluminum and polonium. They have both metallic and nonmetallic properties.

These include: B, Si, Ge, As, Sb, Te, and Po

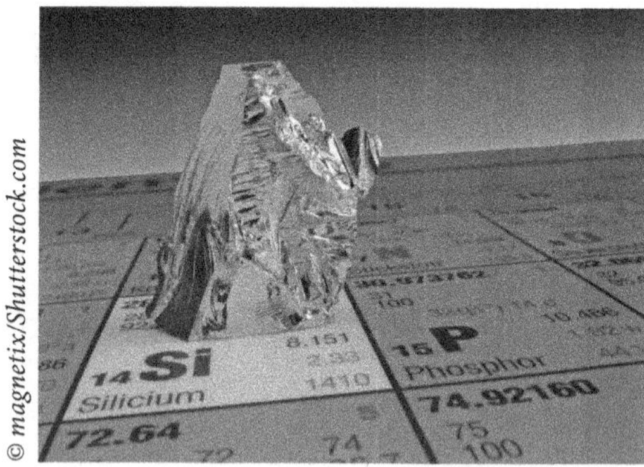

Example: Silicon used in transistors, solar cells, silicones. Silica (sand) used for glass, computer chips, concrete and bricks.

Sources:

[1]Bernard Jaffe, Crucibles: *The Story of Chemistry from Ancient Alchemy to Nuclear Fission* (New York: Simon and Schuster, 1984)

[2]Ibid., 276.

[3]Ibid.

[4]Ibid., 283–284.

[5]Ibid., 306.

Others:

http://www.brooklyn.cuny.edu/bc/ahp/FonF/Dalton.html

http://www.dpgraph.com/janine/twins.html

http://www.rsc.org/images/essay1_tcm18-17763.pdf

CHEMISTRY UNIT 3 PRACTICE PROBLEMS

HISTORY OF THE ATOM

What does the atom look like?

Early Discoveries:

J. J. Thomson   Ernest Rutherford   Democritus   John Dalton   James Chadwick   Robert Millikan

Marie Curie   Dmitri Mendeleev

Gold foil experiment   Alpha particles through nitrogen   Cathode ray tube   Plum Pudding Model

Oil Drop experiment   Discovered radium   Arranged Elements based upon Atomic Weight (Mass)

The Greek's Atom, ~400 BC © Georgios Kollidas/ Shutterstock.com	**The Atomic Theory, 1803** © Georgios Kollidas/ Shutterstock.com
Periodic Table, 1869 © Olga Popova/ Shutterstock.com	**The Electron, 1897** © felichy/ Shutterstock.com
Radioactive Elements, 1898 © catwalker/ Shutterstock.com	**The Electron, 1909** © catwalker/ Shutterstock.com
The Proton, 1902 © rook76/ Shutterstock.com	**The Neutron, 1932** © general-fmv/ Shutterstock.com
The Atom, 1904 © Iaryna Turchyniak/ Shutterstock.com	**The Atom, 1911** © Iaryna Turchyniak/ Shutterstock.com

ISOTOPES

Two Atoms of Lithium:

Element	Number of Protons	Number of Neutrons	Number of Electrons	Atomic Number	Mass Number	Average Atomic Mass
lithium		4				
carbon					13	
	47	61				
lead					207	
	20	19				
tantalum					181	
radium	88				225	
samarium					151	

Creating a Cation:

Creating an Anion:

QUESTION GROUP #1

Directions and/or Common Information:

Isotope Name	Isotope Mass (amu)	Percentage
Silver-107	107	51.86
Silver-109	109	remainder

Find the missing percentage.

1. ______________

Find the average atomic mass of an atom of silver.

2. ______________

QUESTION GROUP #2

Directions and/or Common Information: Silicon has three naturally occurring isotopes.

Isotope Name	Isotope Mass (amu)	Percentage
Silicon-28	28	92.21
Silicon-29	29	4.70
Silicon-30	30	3.09

Now, find the average atomic mass of an atom of silicon

1. ______________

QUESTION GROUP #3

Directions and/or Common Information: Iron has four isotopes.

Isotope Name	Isotope Mass (amu)	Percentage
Iron-54	5.90%	54
Iron-56	91.72%	56
Iron-57	2.10%	57
Iron-58	0.280%	58

Estimate the average mass.

1. ______________

Calculate the average atomic mass of iron.

2. ______________

M&M® CANDY ISOTOPE ACTIVITY

This is Isotopes for the Element: Mm

Red Isotope	Blue Isotope	Yellow Isotope	Green Isotope	Brown Isotope	Orange Isotope	
Ave. Mass (Red) 52 Amu	**Ave. Mass (Blue) 53 Amu**	**Ave. Mass (Yellow) 55 Amu**	**Ave. Mass (Green) 56 Amu**	**Ave. Mass (Brown) 54 Amu**	**Ave. Mass (Orange) 57 Amu**	
Amount (red)	Amount (blue)	Amount (yellow)	Amount (green)	Amount (brown)	Amount (orange)	**TOTAL AMOUNT OF M&Ms (in package)**
Relative abundance (red)	Relative abundance (blue)	Relative abundance (yellow)	Relative abundance (green)	Relative abundance (brown)	Relative abundance (orange)	**RELATIVE ABUNDANCE = amount/total**
Percent abundance (red)	Percent abundance (blue)	Percent abundance (yellow)	Percent abundance (green)	Percent abundance (brown)	Percent abundance (orange)	**PERCENT ABUNDANCE = rel. abundance × 100**
Relative mass (red)	Relative mass (blue)	Relative mass (yellow)	Relative mass (green)	Relative mass (brown)	Relative mass (orange)	**RELATIVE MASS = rel. abundance × ave. mass**

Calculate the Average Atomic Mass of the Element: Mm

Average atomic mass = total of relative masses

PERIODIC TABLE

HISTORY OF THE ATOM

1. How did Dalton come up with the Atomic Theory?

2. Discuss the two compounds (I) carbon monoxide (CO) and (II) carbon dioxide (CO_2). Using Dalton's atomic theory, explain how these two compounds (and their elements) exist.

3. Take the compounds (I) copper oxide (Cu_2O) and (II) copper oxide: CuO. If the mass ratio in Cu_2O is 12.0 grams Cu and 6.0 grams O; suggest the mass of O possible in CuO if Cu remains 12.0 grams.

4. Name three researchers who worked with cathode ray tubes.

5. Ernest Rutherford discovered protons as being positive hydrogen atoms. Is this correct?

6. In 1897, J. J. Thomson discovered electrons, in 1911 Ernest Rutherford discovered protons, and in 1932 James Chadwick discovered neutrons. Why do you think these subatomic particles were discovered in this order?

7. Why is Rutherford's model of the atom better than Thomson's?

8. Who helped Rutherford discover the nucleus of an atom by an experiment famously known as the gold foil experiment?

9. Which one of Dalton's major conclusions explained the law of conservation of mass?
   a. Matter is composed of small, indivisible particles called atoms.
   b. Atoms of the same element are identical and have the same properties.
   c. Compounds are composed of atoms of different elements combined in small whole-number ratios.
   d. Atoms of the same element have the same mass.
   e. Reactions are merely the rearrangement of atoms into different combinations.

10. Who contributed the following: The charge of the electron was determined by performing the oil drop experiment.

11. Who contributed to the discovery of the neutron?

12. Who contributed the following: The charge/mass ratio for 20 different metals in the cathode ray tube was the same, which suggested that electrons are present in all kinds of matter.

13. Taken by itself the fact that 8.0 grams of oxygen and 1.0 grams of hydrogen combine to give 9.0 grams of water demonstrate what natural law?
    a. conservation of energy
    b. conservation of mass
    c. periodicity
    d. atomic theory
    e. atomic number

14. Who discovered the nucleus in the atom?

STRUCTURE OF THE ATOM—PROTONS, NEUTRONS, ELECTRONS

1. The mass number of an atom is equal to _____.
    a. the number of electrons in the nucleus
    b. the number of protons in the nucleus
    c. the number of neutrons in the nucleus
    d. the sum of the protons and neutrons in the nucleus

2. The number of protons found in an atom is equal to _____.
    a. the number of electrons in the nucleus
    b. the atomic number of the element
    c. the number of neutrons in the nucleus
    d. the sum of the protons and neutrons in the nucleus

3. The number of neutrons found in an atom is equal to _____.
    a. the number of electrons in the nucleus
    b. the mass number of the atom
    c. the number of protons in the nucleus
    d. the difference between the mass number and the atomic number in the atom

4. Complete the following table.

Atom	Charge	Ion	#p⁺	#e⁻
Sr	+2	Sr^{2+}	38	38–2 = 36
O		O^{2-}		
C	+4			
Fe	+3			
Sb	–3			
Br		Br^{-}		

5. How many protons, neutrons, and electrons does the element $^{65}_{30}Zn^{2+}$ contains:

6. What is the total number of subatomic particles in $^{44}_{21}Sc^{3+}$?

7. Which of the following is an isotope of $^{204}_{82}Pb$?
   a. $^{199}_{81}Tl$
   b. $^{202}_{80}Hg$
   c. $^{212}_{82}Pb$
   d. $^{206}_{80}Hg$
   e. $^{197}_{81}Tl$

8. The isotopes of the same element have _____.
   a. different number of protons
   b. different number of neutrons
   c. different atomic numbers
   d. different number of electrons

9. Find the number of the neutrons and electrons for lead-208 (this is the mass number).

10. Complete the following table.

Element	Atomic Number	Mass Number	#p⁺	#n⁰	#e⁻	Isotopic Notation
K	19	39	19	20	19	$^{39}_{19}K$
Br		80				
Ar				22		
Cr		52			22	
Si				14	10	

11. Which one of the following statements about the isotopes of a given element is correct?
    a. The atoms have the same number of protons but differing numbers of neutrons.
    b. The atoms have the same number of neutrons but differing numbers of electrons.
    c. The atoms have the same number of electrons but differing numbers of protons.
    d. The atoms have the same number of neutrons but differing numbers of protons.
    e. The atoms have the same number of protons but differing numbers of electrons.

ISOTOPES & ATOMIC MASSES

1. How are isotopes and average atomic masses related?

2. Write down the correct atomic mass for each of the following elements.
   a. tin
   b. antimony
   c. argon
   d. lithium

3. Naturally occurring copper is 69.09% ^{63}Cu (62.96 amu). The only other isotope is ^{65}Cu (64.96 amu). What is the atomic mass of copper?

4. The average atomic mass (amu) of each element is determined by _____.
   a. adding the number of protons and neutrons in any atom of the element
   b. adding the number of protons and electrons in any atom of the element
   c. adding the number of neutrons and electrons in any atom of the element
   d. adding the number of protons in any atom of the element
   e. None of the above are true.

5. An element has an atomic mass of 35.5. It probably contains a mixture of _____.
   a. isomers
   b. allotropes
   c. radioactive atoms
   d. isotopes
   e. optics

6. Argon in nature consists of the following isotopes:

Isotope	Atomic Mass (amu)	Percent Abundance
Argon-36	35.968 amu	0.337%
Argon-38	37.963 amu	0.063%
Argon-40	39.962 amu	99.600%

Calculate the average atomic mass of argon.

THE PERIODIC TABLE

1. How does the modern periodic table differ from Mendeleev's periodic table?

2. Fill in the blank periodic table as complete as possible.

3. Define position and symbol in Periodic Table of the Iron.
   a. period 4, group VIIIb, symbol Fe
   b. period 5, group VIIIa; symbol I
   c. period 2, group VIIa, symbol F
   d. period 6; group VIIIb; symbol Ir

4. Which of the following elements is a noble gas?
   a. Li
   b. Mg
   c. Ni
   d. Br
   e. Rn

5. What is the name for the elements in Group 2A?

6. Element No. 15 is _____.
   a. an alkaline earth metal
   b. a transition metal
   c. a nonmetal
   d. a noble gas
   e. a metalloid

7. Which of the following elements is a nonmetal?
   a. Bi
   b. Ge
   c. B
   d. Se
   e. Si

8. Element No. 36 is _____.
   a. an alkaline earth metala transition metal
   b. a nonmetal
   c. a noble gas
   d. a metalloid

9. All of the following nonmetals exist as diatomic molecules except _____.
   a. F
   b. H
   c. N
   d. I
   e. Kr

10. The chemical properties of selenium would be most similar to _____.
    a. As
    b. Br
    c. S
    d. P
    e. I

11. Which pair of elements could be called metalloids?
    a. Li and Na
    b. Cu and Ag
    c. F and Cl
    d. Si and Ge
    e. Na and Mg

12. Element No. 26 is _____.
    a. an alkaline earth metal
    b. transition metal
    c. a nonmetal
    d. a noble gas
    e. a metalloid

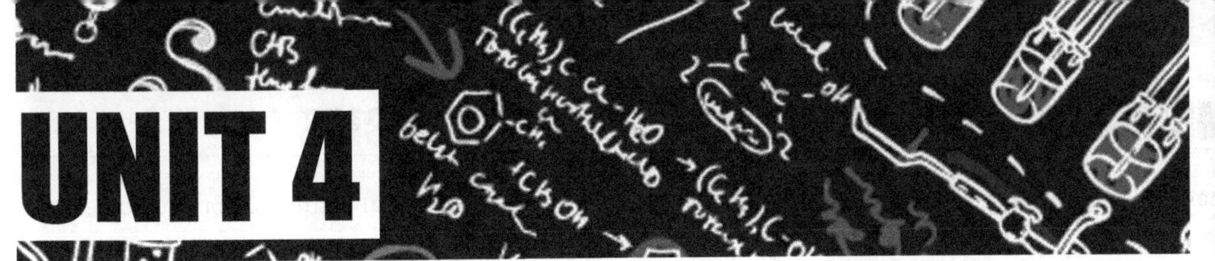

NAME THAT COMPOUND (NOMENCLATURE)

OBJECTIVES FOR UNIT 4

- To name or give the formula of binary molecular compounds.
- To name or give the formula of ionic compounds.
- To recognize the formula for a hydrate.
- To recognize and name Arrhenius acids and bases.

CHEMICAL FORMULAS FOR MOLECULES AND PARTICLES

As we have seen with elements, a shorthand notation is used to write chemical symbols. For molecules and particles, this shorthand notation involves writing chemical formulas.

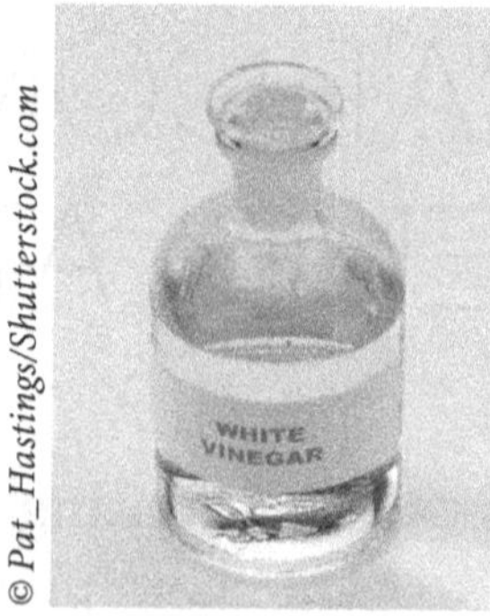

© Pat_Hastings/Shutterstock.com

Acetic acid is the acid found in vinegar.

The formula $HC_2H_3O_2$ indicates 2 atoms of C, 4 atoms of H, 2 atoms of O

Sometimes acetic acid is written as CH_3COOH. This is called its structural formula.

© bouzou/Shutterstock.com

Sulfur hexafluoride is a heavy gas often used in high electrical environments to prevent sparks from passing. The formula SF_6 indicates 1 atom S, 6 atoms F.

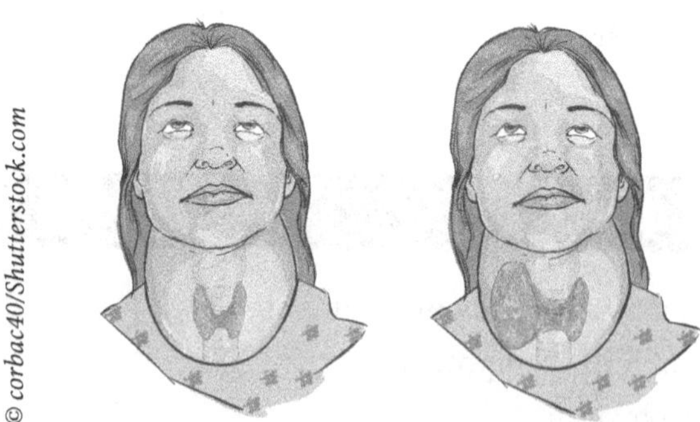
© corbac40/Shutterstock.com

Potassium iodide was added to table salt (iodized salt) to prevent a deficiency of iodine, which can lead to a condition called goiter (an enlargement of the thyroid gland).

The formula KI indicates 1 atom K, 1 atom I.

INTRODUCTION TO NOMENCLATURE

In Unit 3 we learned about atoms and how different atoms make up different elements. In chemistry we consider atoms as being the building blocks of all the matter that we study. Just like Legos® are put together to make different shapes, atoms combine to make different material. In order to understand how atoms do this, we need to learn about chemical bonding.

CHEMICAL BOND

A chemical bond (ionic and molecular): Because atoms carry electron(s), and proton(s), there is attractiveness, not just to themselves but to other atoms. Most atoms have an attraction for one another, similar to how a magnet works. Atoms combine with each other by sharing or transferring electrons.

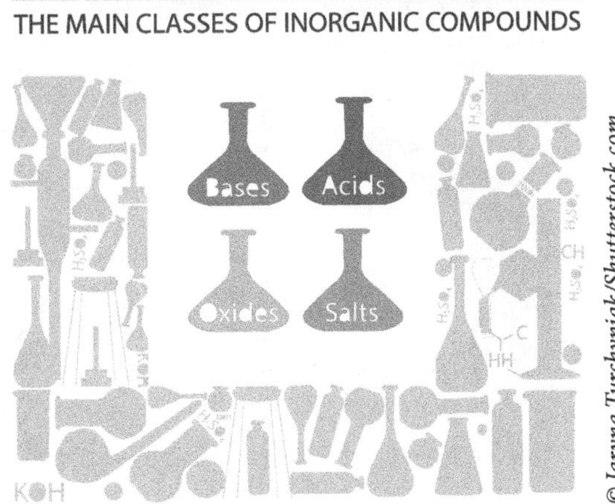

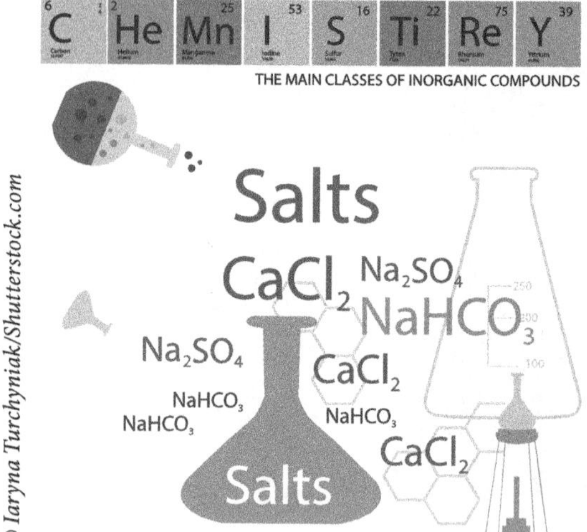

IONIC BONDS

When the attraction between atoms involves the transfer of one or more electrons, we call this **ionic bonding**.

These bonds involve metals (or ammonium) with nonmetals (or negative polyatomics).

An ionic bond results when an electropositive (usually metal) meets an electronegative (usually a nonmetal), creating an electrostatic attraction.

Examples:

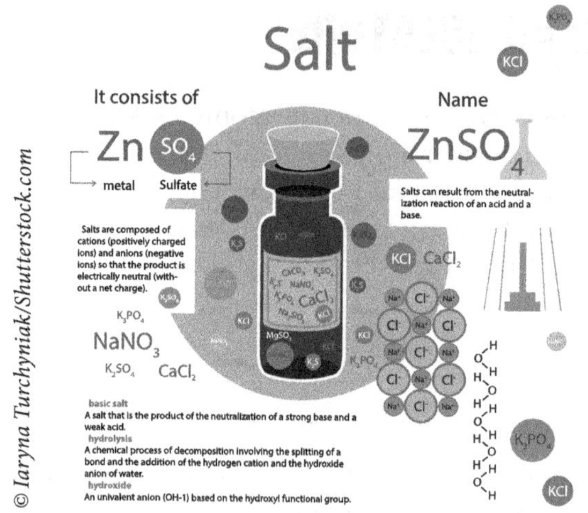

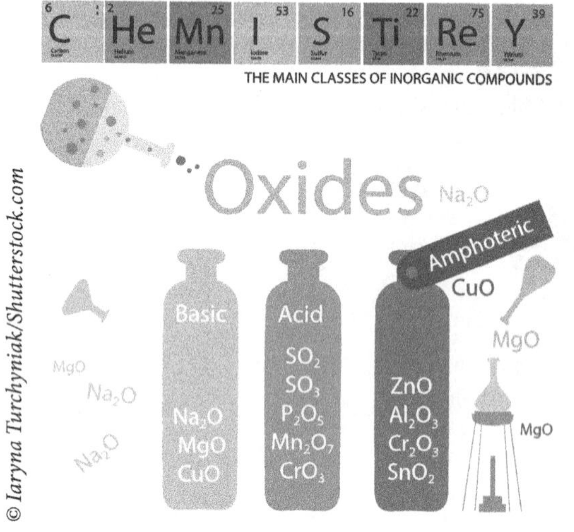

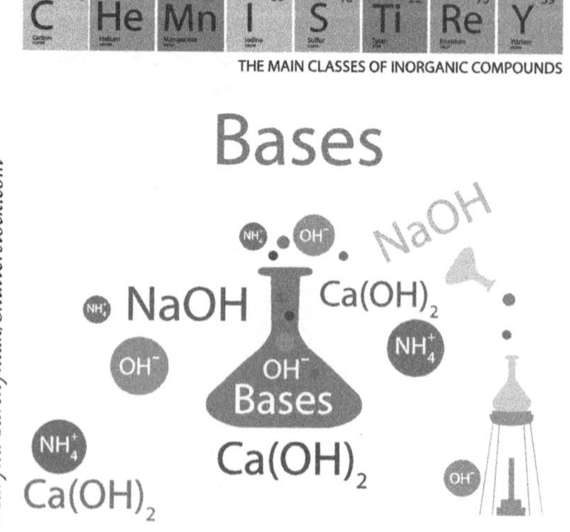

Note: compounds that form ionic bonds are called particles, salts, or formula units.

MOLECULAR BONDS

When the attraction between atoms involves the sharing of electrons, we call this **molecular bonding**. Molecular bonds are also known as COVALENT BONDS.

These bonds involve a bond between a nonmetal + nonmetal(s)

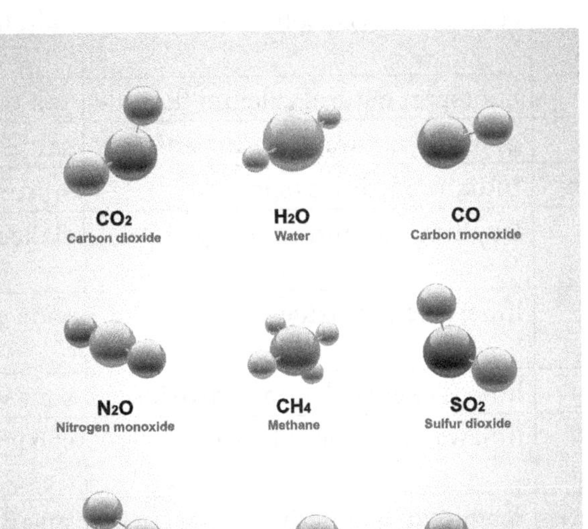

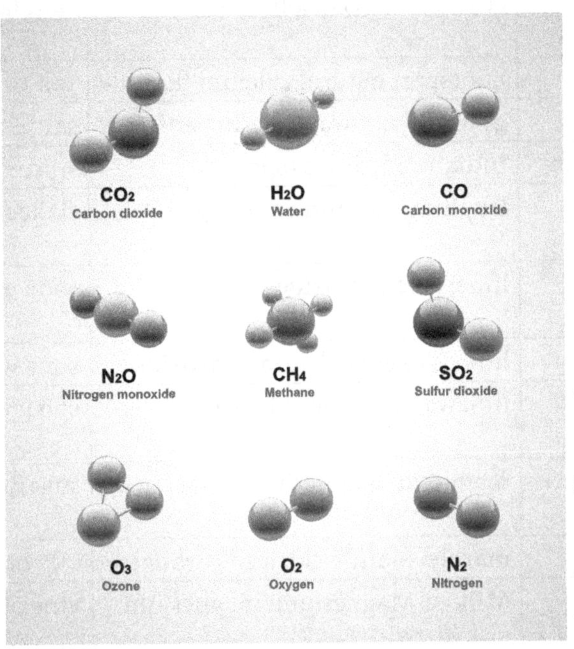

Example in molecules:

Note: compounds that form molecular bonds are called molecules.

NOMENCLATURE

Nomenclature is a way of naming compounds (both ionic and molecular). There are two types of nomenclature: common name and systematic name.

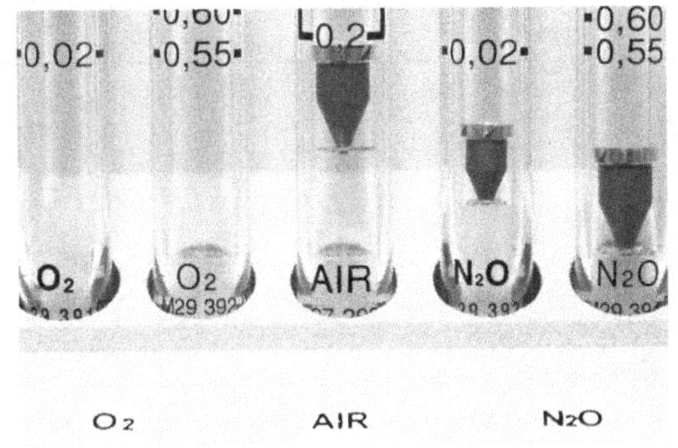

Dinitrogen monoxide is an anesthetic. Known as laughing gas, the formula is N_2O. If someone accidentally used nitrogen dioxide (NO_2) instead of dinitrogen monoxide, the patient might go unconscious. (Nitrogen dioxide is the brownish haze in polluted air.)

SOME COMMON NAMES

acetone: dimethyl ketone	diamond: carbon crystal	quicksilver: mercury
alcohol, grain: ethyl alcohol	emery powder: impure aluminum oxide	rock salt: sodium chloride
alcohol, wood: methyl alcohol	Epsom salts: magnesium sulfate	rubbing alcohol: isopropyl alcohol
alum: aluminum potassium sulfate	ethanol: ethyl alcohol	salt, table: sodium chloride
asbestos: magnesium silicate	fluorspar: natural calcium fluoride	salt of tartar: potassium carbonate
aspirin: acetylsalicylic acid	gypsum: natural calcium sulfate	saltpeter: potassium nitrate
baking soda: sodium bicarbonate	Indian red: ferric oxide	silica: silicon dioxide
banana oil (artificial): isoamyl acetate	laughing gas: nitrous oxide	slaked lime: calcium hydroxide
black ash: crude sodium carbonate	lime: calcium oxide	soda ash: sodium carbonate
black lead: graphite (carbon)	lime, slaked: calcium hydroxide	sour water: dilute sulfuric acid
bleaching powder: calcium hypochlorite	limewater: calcium hydroxide (aq)	spirit of wine: ethyl alcohol
bone ash: crude calcium phosphate	magnesia: magnesium oxide	sugar, table: sucrose
borax: sodium borate	marble: mainly calcium carbonate	talc or talcum: magnesium silicate
brimstone: sulfur	Milk of Magnesium: magnesium hydroxide	vinegar: impure dilute acetic acid
brine: sodium chloride (aq)	muriatic acid: hydrochloric acid	vitamin C: ascorbic acid
caustic lime: calcium hydroxide	oil of vitriol: sulfuric acid	vitriol: sulfuric acid
caustic potash: potassium hydroxide	oil of wintergreen: methyl salicylate	washing soda: sodium carbonate
caustic soda: sodium hydroxide	plaster of Paris: calcium sulfate	water glass: sodium silicate
chalk: calcium carbonate	potash: potassium carbonate	
cream of tartar: potassium bitartrate	quicklime: calcium oxide	

WRITING FORMULA/NAMING IONIC COMPOUNDS (PARTICLES, SALTS, FORMULA UNITS)

Because an ionic compound's bond involves the transfer of electron(s) from one atom to another, we need to look the atoms that offer the electron(s) and the atoms that accept the electron(s). These atoms are called ions.

IONS

Ions are atoms that carry a charge because they lose or gain electrons.

There are two types of ions:

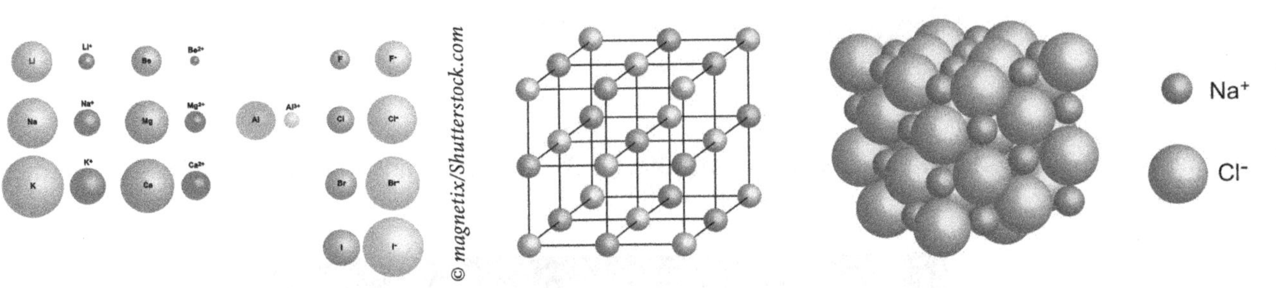

Cations are ions that carry a positive charge because they are most stable when they lose electrons. In an ionic bond, the metal atoms form the cations (along with the positive polyatomic ammonium).

EXAMPLES OF CATIONS

Copper ion	
Copper (I) Cuprous	Cu^+
Copper (II) Cupric	Cu^{2+}
Ammonium	NH_4^+

Anions are ions that carry a negative charge because they are most stable when they gain electrons. In an ionic bond, the nonmetal atoms form the anions, along with the negative polyatomics. (Refer to in class practice problems for complete table).

EXAMPLES OF ANIONS

Phosphide ion	P^{3-}
hypochlorite	ClO^-
chlorite	ClO_2^-
chlorate	ClO_3^-
perchlorate	ClO_4^-

The charges on cations and anions are referred to as the oxidation numbers. To write the formula for a compound, the sum of the charges (oxidation numbers) must be equal to zero. The ionic compounds will be written neutral with the cation charge balancing out the anion charge.

Cations and Anions that involve only one atom are called monoatomic ions.

Cations and Anions that involve more than one atom are called polyatomic ions.

Calcium, Ca^{2+} is a cation that bonds to an anion sulfate, SO_4^{2-} to make calcium sulfate, $CaSO_4$, an ingredient of wallboard.

Magnesium oxide is used in 4th of July sparklers. Its formula is: MgO.

The oxidation numbers are as follows: Mg^{2+} O^{2-},

When the formula is written as MgO, the sum of the oxidations numbers equal zero.

Some more examples of writing formulas of ionic compounds:

Compound	Cation(s)	Anion(s)
Fe_2O_3	iron: 2 Fe^{3+}	oxygen: 3 O^{2-}
$CaCl_2$	calcium: Ca^{2+}	chlorine: 2 Cl^-
Li_2S	lithium: 2 Li^+	sulfur: S^{2-}
SnO	tin: Sn^{2+}	oxygen: O^{2-}
$Pb(NO_3)_2$	lead: Pb^{2+}	nitrate: 2 NO_3^-
$(NH_4)_2SO_4$	ammonium: 2 NH_4^+	sulfate: SO_4^{2-}

TO NAME SIMPLE IONIC COMPOUNDS

(metal stable with only one possible charge)

- Name the cation (metal) first or the positive ammonium polyatomic
- Name the anion (nonmetal) second, changing its ending to *-ide*, or the negative polyatomic

  Note: If the anion is a polyatomic, do not change the ending.

Antacids

Stomach
Antacid
Pyloric sphincter
Stomach acid
Duodenum

© joshya/Shutterstock.com

One common antacid is called Milk of Magnesia which is magnesium hydroxide. The hydroxide in this formula is a polyatomic ion.

The formula: $Mg(OH)_2$

Some more examples of naming simple ionic compounds:

Formula	Metal	Nonmetal	Compound Name
KI	potassium	iodine	potassium iodide
Li_2S	lithium	sulfur	lithium sulfide
Al_2O_3	aluminum	oxygen	aluminum oxide
Mg_3N_2	magnesium	nitrogen	magnesium nitride
$CaCO_3$	calcium	carbonate	calcium carbonate
$Al_2(Cr_2O_7)_3$	aluminum	dichromate	aluminum dichromate

TO NAME COMPLEX IONIC COMPOUNDS

(metal stable with more than one possible charge)

- Name the cation using roman numerals (in parenthesis) or use ion name (derived from the name of the metal or Latin root) to indicate which oxidation number of the cation is being used.
- Still name the anion (nonmetal) *-ide*.

  Note: If the anion is a polyatomic, do not change the ending!!

© TTstudio/Shutterstock.com

Upper Michigan at one time mined most of America's copper. When copper reacts with the atmosphere, it produces copper (II) oxide.

Some more examples of naming complex ionic compounds:

Formula	Cation	Anion	Compound Name
CuO	copper (II), Cu^{2+} or cupric, Cu^{2+}	oxide, O^{2-}	copper (II) oxide or cupric oxide
$SnCl_2$	tin (II), Sn^{2+} or stannous, Sn^{2+}	chloride, Cl^-	tin (II) chloride or stannous chloride
Fe_2O_3	iron (III), Fe^{3+} or ferric, Fe^{3+}	oxide, O^{2-}	iron (III) oxide or ferric oxide
PbC_2O_4	lead (II) Pb^{2+} or plumbous, Pb^{2+}	oxalate, $C_2O_4^{2-}$	lead (II) oxalate or plumbous oxalate
$CrPO_4$	chromium (III), Cr^{3+} or chromic, Cr^{3+}	phosphate, PO_4^{3-}	chromium (III) phosphate or chromic phosphate

HYDRATES

Many naturally existing minerals contain ionic compounds or salts. These compounds are found with differing properties and colors. Some of these salts are considered precious because of their colors. Such salts fall under the category of hydrated compounds. Hydrated compounds are ionic compounds that have bonded to the water molecule.

TO WRITE THE FORMULA FOR HYDRATES

- Write the ionic compound following ionic nomenclature methods.
- The prefixes before the word hydrate indicate the quantity of water molecules bonded.

$CuSO_4{\bullet}5H_2O$ is a very beautiful blue mineral—copper(II) sulfate pentahydrate

Prefixes: mono, di, tri, tetra, penta, hexa, hepta, octa, nona, deca

TO NAME HYDRATES

- Name the ionic compound following ionic nomenclature methods.
- Use a to separate between the ionic compound and the water molecule
- Use prefixes to indicate the quantity of water molecules bonded with the word hydrate.

Ferric oxide trihydrate is a hematite mineral. $Fe_2O_3{\bullet}3H_2O$

WRITING FORMULA/NAMING MOLECULAR COMPOUNDS (MOLECULES)

TYPE 1: NONACID MOLECULAR COMPOUNDS—COMPOSED OF TWO NONMETAL ELEMENTS

Nonacid molecular compounds involve the sharing of electrons (not always an equal sharing) between two nonmetals. Because these atoms share electrons, we do not look at charges.

Note: it should be mentioned that many metalloids also form nonacid molecular compounds.

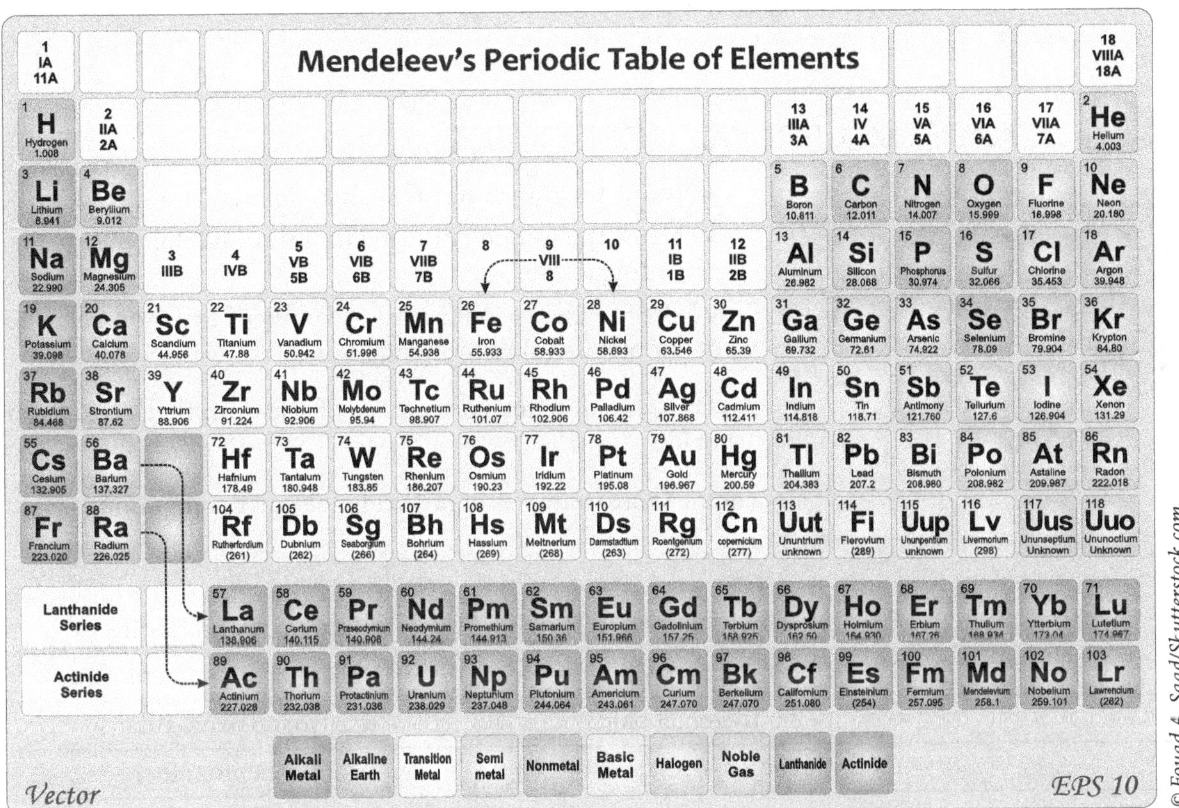

On the above periodic table the only elements of focus are those on the stairstep line, those to the right of the stairstep line, and hydrogen.

TO WRITE THE FORMULA FOR NONACID MOLECULAR COMPOUNDS

- Write the symbol for the two nonmetals (or metalloids).
- The prefixes before the nonmetals indicate the quantity of the nonmetals (or metalloid) bonded.

PREFIXES

MONO	DI	TRI	TETRA	PENTA	HEXA	HEPTA	OCTA	NONA	DECA
1	3	3	4	5	6	7	8	9	10

Tetraphosphorus trisulfide: P_4S_3 is the chemical on the head of a strike-match.

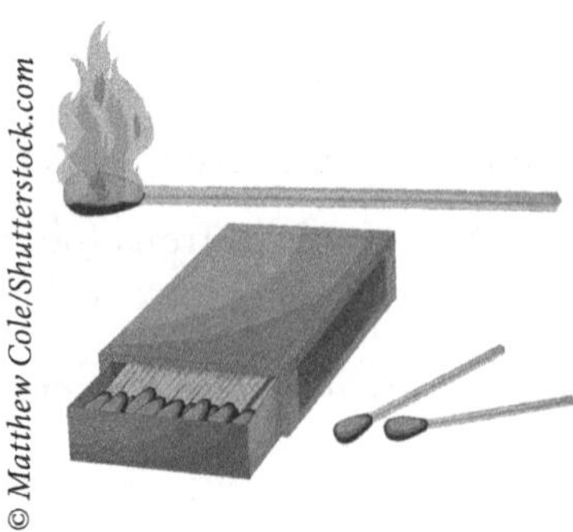
© Matthew Cole/Shutterstock.com

TO NAME NONACID MOLECULAR COMPOUNDS

© designer491/Shutterstock.com

- Name the full name of the first nonmetal (or metalloid).
- Name the second nonmetal (or metalloid) but change the ending to -ide.
- Use prefixes to indicate the quantity of each nonmetal atom (or metalloid) in the bond.

SeO_2 is one of many compounds of selenium that exist. Selenium dioxide

Some more examples of nonacid molecular compounds:

Compound	Atoms	Name
PCl_3	1 phosphorus, 3 chlorine	phosphorus trichloride
PCl_5	1 phosphorus, 5 chlorine	phosphorus pentachloride
N_2O_5	2 nitrogen, 5 oxide	dinitrogen pentoxide
S_2O_7	2 sulfur, 7 oxide	disulfur heptoxide

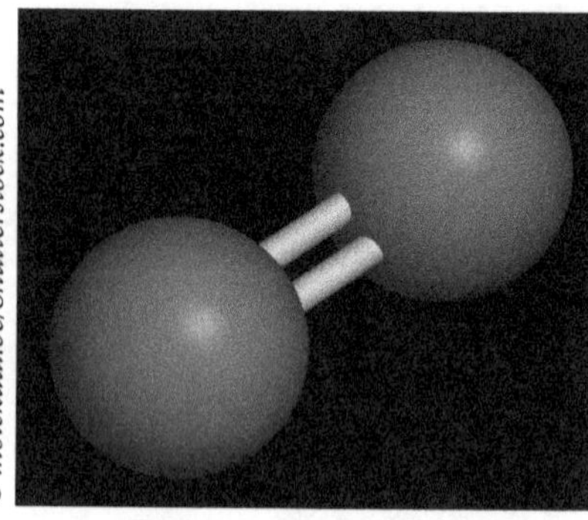
© molekuul.be/Shutterstock.com

Note: If the first element is just one atom, often mono is dropped; also look at the vowel ending.

Review the diatomic molecules because these molecular compounds use their element names:

H_2, O_2, N_2, F_2, Cl_2, Br_2, I_2

H_2O is separated into H_2 (hydrogen) and O_2 (oxygen).

TYPE 2: ACID MOLECULAR COMPOUNDS—COMPOSED OF H BONDED TO NONMETAL ELEMENT(S)

Acid molecules are unique in that they share electrons in their bond but when placed into a water environment they transfer electrons. The transfer takes place from the hydrogen atom(s) that lose an electron to the nonmetal(s). One property of acids is a sour taste.

NONOXOACID MOLECULES

Nonoxoacids are molecules with H bonded to a single nonmetal

TO WRITE THE FORMULA FOR NONOXOACID MOLECULAR COMPOUNDS

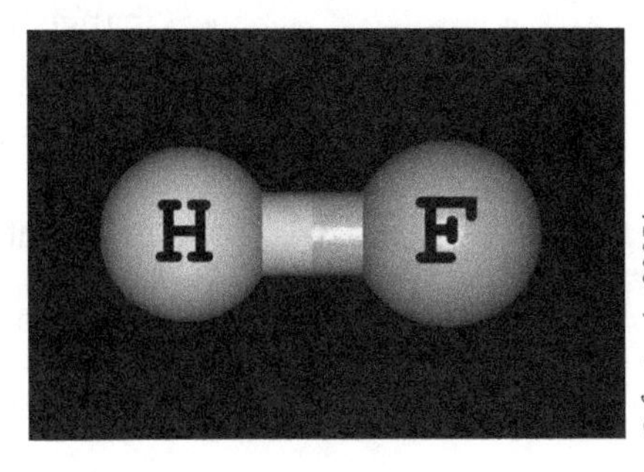

© foxterrier2005/ Shutterstock.com

- The cation is hydrogen as H^+, the anion is a nonmetal (located to the left of the stairstep line on the periodic table).
- Write the chemical formula so the charges are neutral.

One of the most dangerous acids is hydrofluoric acid.

H+ bonded to F- forms the compound HF

TO NAME NONOXOACID MOLECULAR COMPOUNDS

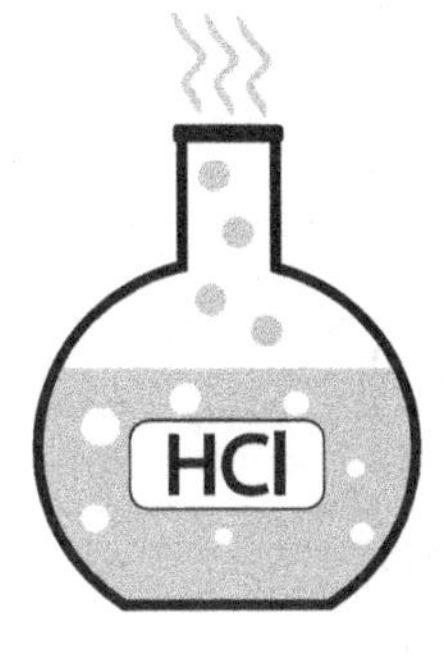

© konstantinks/Shutterstock.com

- Write the prefix hydro-
- Change the ending -ide to -ic acid

Stomach acid is the strong acid: HCl. It is H^+ bonded to F^- which is fluoride

Hydrofluoric acid

Some more examples of nonoxoacid molecular compounds:

Compound	Anion	Acid Name
HCl	chloride, Cl^-	hydrochloric acid
HF	fluoride, F^-	hydrofluoric acid
H_2S	sulfide, S^{2-}	hydrosulfuric acid

Note: sulfide does not change to hydrosulfic acid; it is properly written: hydrosulfuric acid.

OXOACID MOLECULES

Oxoacids are molecules with H bonded to a negative polyatomic

TO WRITE THE FORMULA FOR OXOACID MOLECULAR COMPOUNDS

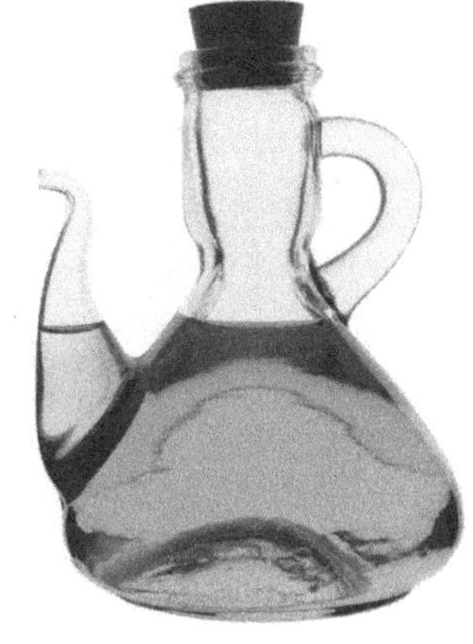

© Nigel Paul Monckton/Shutterstock.com

- The cation is hydrogen as H^+; the anion is a negative polyatomic.
  If the acid ends in -ous acid it is H bonded to a polyatomic ending in -ite
  If the acid ends in -ic acid it is H bonded to a polyatomic ending in -ate
- Write the chemical formula so the charges are neutral.

The acid contained in vinegar is acetic acid

H^+ bonded to acetate which is $C_2H_3O_2^-$

$HC_2H_3O_2$

TO NAME NONOXOACID MOLECULAR COMPOUNDS

- Never write the prefix hydro-
- Change the ending
  -ite ending changes to -ous acid
  -ate ending changes to -ic acid

Soda pop contains the acid H_2CO_3

2 H^+ bonds to CO_3^{2-} which is the polyatomic ion carbonate

Carbonic acid

Some more examples of oxoacid molecular compounds:

Compound	Anion	Acid Name
HClO	hypochlorite, ClO^-	hypochlorous acid
$HClO_2$	chlorite, ClO_2^-	chlorous acid
$HClO_3$	chlorate, ClO_3^-	chloric acid
$HClO_4$	perchlorate, ClO_4^-	perchloric acid
H_2SO_3	sulfite, SO_3^{2-}	sulfurous acid
H_3PO_4	phosphate, PO_4^{3-}	phosphoric acid

Note: sulfite does not change to sulfous acid; it is properly written sulfurous acid. Phosphate does not change to phosphic acid; it is properly written phosphoric acid.

Common Cations	
Aluminium ion	Al^{3+}
Barium ion	Ba^{2+}
Beryllium ion	Be^{2+}
Bismuth	Bi^{3+}
Cadmium ion	Cd^{2+}
Calcium ion	Ca^{2+}
Cesium ion	Cs^{+}
Chromium ion	
Chromium (II)	Cr^{2+}
Chromous	
Chromium (III)	Cr^{3+}
Chromic	
Cobalt ion	
Cobalt (II)	Co^{2+}
Cobaltous	
Cobalt (III)	Co^{3+}
Cobaltic	
Copper ion	
Copper (I)	Cu^{+}
Cuprous	
Copper (II)	Cu^{2+}
Cupric	
Francium	Fr^{+}
Gold ion	
Gold (I)	Au^{+}
Aurous	
Gold (III)	Au^{3+}
Auric	
Iron ion	
Iron (II)	Fe^{2+}
Ferrous	
Iron (III)	Fe^{3+}
Ferric	
Lead ion	
Lead (II)	Pb^{2+}
Plumbous	
Lead (IV)	Pb^{4+}
Plumbic	
Lithium ion	Li^{+}

Common Cations	
Magnesium ion	Mg^{2+}
Manganese ion	
Manganese (II)	Mn^{2+}
Manganous	
Manganese (III)	Mn^{3+}
Manganic	
Mercury ion	
Mercury (I)	Hg^{+}
Mercurous	
Mercury (II)	Hg^{2+}
Mercuric	
Nickel ion	
Nickel (II)	Ni^{2+}
Nickelous	
Nickel (III)	Ni^{3+}
Nickelic	
Potassium ion	K^{+}
Radium ion	Ra^{2+}
Rubidium ion	Rb^{+}
Silver ion	Ag^{+}
Sodium ion	Na^{+}
Strontium ion	Sr^{2+}
Tin ion	
Tin (II)	Sn^{2+}
Stannous	
Tin (IV)	Sn^{4+}
Stannic	
Titanium ion	
Titanium (III)	Ti^{3+}
Titanous	
Titanium (IV)	Ti^{4+}
Titanic	
Zinc ion	Zn^{2+}
Polyatomic Cation	
Ammonium	NH_4^{+}
For Acids	
Hydrogen	H^{+}

Common Anions	
Astatide ion	At^-
Bromide ion	Br^-
Carbide ion	C^{4-}
Chloride ion	Cl^-
Fluoride ion	F^-
Hydride ion	H^-
Iodide	I^-
Nitride ion	N^{3-}
Oxide ion	O^{2-}
Phosphide ion	P^{3-}
Selenide ion	Se^{2-}
Sulfide ion	S^{2-}

Polyatomic Anions	
acetate	$C_2H_3O_2^-$
arsenate	AsO_4^{3-}
borate	BO_3^{3-}
bromate	BrO_3^-
bromite	BrO_2^-
carbonate	CO_3^{2-}
chlorate	ClO_3^-
chlorite	ClO_2^-
chromate	CrO_4^{2-}
citrate	$C_6H_5O_7^{3-}$
cyanate	OCN^-
cyanide	CN^-
dichromate	$Cr_2O_7^{2-}$
dihydrogen phosphate	$H_2PO_4^-$
ferricyanide	$Fe(CN)_6^{3-}$

Polyatomic Anions	
ferrocyanide	$Fe(CN)_6^{4-}$
hydrogen carbonate (bicarbonate)	HCO_3^-
hydrogen phosphate (biphosphate)	HPO_4^{2-}
hydrogen sulfate (bisulfate)	HSO_4^-
hydrogen sulfide (bisulfide)	HS^-
hydrogen sulfite (bisulfite)	HSO_3^-
hydroxide	OH^-
hypochlorite	ClO^-
iodate	IO_3^-
nitrate	NO_3^-
nitrite	NO_2^-
oxalate	$C_2O_4^{2-}$
perbromate	BrO_4^-
perchlorate	ClO_4^-
periodate	IO_4^-
permanganate	MnO_4^-
phosphate	PO_4^{3-}
selenate	SeO_4^{2-}
silicate	SiO_3^{2-}
sulfate	SO_4^{2-}
sulfite	SO_3^{2-}
tellurate	TeO_6^{6-}
thiocyanate	SCN^-
thiosulfate	$S_2O_3^{2-}$

CHEMISTRY UNIT 4 PRACTICE PROBLEMS

IONIC NOMENCLATURE

Ionic bond (form ____):

Ionic bonds are bonds between: ____________________ Exception: ____________________

KBr	
$BaSO_4$	
Cu_2O	
$Fe_2(SO_4)_3$	
K_3PO_4	
Al_2O_3	
NH_4Cl	
CuOH	
$Ca(C_2H_3O_2)_2$	
$ZnSO_3$	
$SnCl_2$	

Barium nitrate	
Potassium iodide	
Iron (III) bromide	
Copper (I) phosphate	
Zinc sulfite	
Calcium hydroxide	
Tin (IV) sulfide	
Magnesium carbonate	
Ammonium nitrate	
Aluminum sulfate	
Tin (II) bromide	
Sodium bicarbonate	

Ionic Nomenclature

ammonium sulfide	nickel (II) iodide	sodium nitrate	mercurous oxide
cupric bromide	lead(II) chlorite	aluminum sulfate	potassium nitrate
iron(II) bisulfate	ferrous carbonate	magnesium nitrate	lead(II) phosphate
iron(III) chromate	iron(II) chromate	cupric hydroxide	copper(II) hydroxide
calcium fluoride	cuprous carbonate	nickel (II) nitrate	chromic acetate
silver cyanide	calcium chlorate	$(NH_4)_2SO_3$	$(NH_4)_2O$
$ZnSO_4$	$Al(ClO_4)_3$	$SnCl_2$	$Zn(HCO_3)_2$
$FeCl_3$	Na_3PO_4	Ag_2S	AgClO
$Mg(OH)_2$	$(NH_4)_3PO_4$	$(NH_4)_2CO_3$	$Fe(ClO_2)_2$
$Ni(C_2H_3O_2)_2$	K_2S	Na_2CrO_4	$SnBr_4$
$Cr(HSO_4)_3$	Li_2CrO_4	$KMnO_4$	$Mg(HSO_4)_2$
$AgClO_4$	$Fe_3(PO_4)_2$	K_3PO_4	

Molecular Nomenclature

nitrogen trihydride	phosphorus pentachloride	phosphorus triiodide
arsenic pentachloride	diphosphorus trioxide	nitrogen dioxide
dinitrogen tetrafluoride	pentaboron nonahydride	tetrasulfur tetranitride
H_2O_2	CS_2	BCl_3
SCl	CCl_4	S_2Cl_2
NO	OF_2	P_4S_{10}

Acid Nomenclature

hydroiodic acid	chlorous acid	chloric acid
sulfurous acid	hydrosulfuric acid	hydroselenic acid
selenic acid	periodic acid	oxalic acid
HF	H_2CO_3	HNO_2
H_6TeO_6	$H_2Cr_2O_7$	HCl
HOCN	$HBrO_2$	$HC_2H_3O_2$

Combined Nomenclature

silver bicarbonate	nitrous acid	barium bisulfate
lead(IV) chlorite	hydrobromic acid	calcium sulfide
triphosphorus heptoxide	hypochlorous acid	copper(I) bisulfate
sulfuric acid	boron triphosphide	cobaltic chlorate
SO_3	$HClO_4$	NH_3
$Ba(BrO_3)_2$	XeF_2	Al_2S_3
H_3PO_4	$Mn_3(PO_4)_2$	H_2Se
H_2SO_3	P_2O_4	HNO_3
$Cr_2(SO_3)_3$	$Ni(ClO_4)_2$	H_3AsO_4
sodium acetate	carbonic acid	tetrasulfur hexaphosphide

STRUCTURE OF CHEMICAL COMPOUNDS

1. C_8H_{18} contains a total of _____ atoms, including _____ atoms C and _____ atoms H.
2. $Al(NO_3)_3$ contains a total of 13 atoms, including 3 atoms N, _____ atoms Al, and _____ atoms O.
3. $(NH_4)_2SO_4$ contains a total of _____ atoms.
4. One particle of $(NH_4)_3PO_4$, ammonium phosphate, contains _____ NH_4^+ ions, _____ PO_4^{3+} ions, _____ atoms N, _____ atoms H, _____ atoms P, _____ atoms O, and _____ total atoms.
5. Which of the following is not a diatomic molecule?
   a. H_2
   b. N_2
   c. K_2
   d. O_2
   e. Cl_2
6. The total number of atoms represented by $Ba(H_2PO_4)_2$ is:
   a. 16
   b. 13
   c. 14
   d. 17
   e. 15
7. Classify each compound as either: M = molecular (two nonmetals) I = Ionic (metal + nonmetal)
   a. $FeCl_2$
   b. N_2O
   c. HBr
   d. CO_2
   e. BaI_2
   f. K_2O
8. List 3 common examples of each of the three types of bonding.
   a. Ionic ______________________________
   b. Covalent ______________________________
   c. Metallic ______________________________
9. Circle each of the following compounds that is ionically bonded:
   CO_2 SiO_2 NaCl HBr CH_4 Rb_2O KF
10. Which of the following compounds are covalently bonded?
    NO_2 PCl_3 NH_3 NaI $CaCl_2$ Al_2O_3 C_6H_6
11. Which of the following pairs of atoms would not be expected to combine to form an ionic compound?
    a. B and H
    b. Na and H
    c. Li and Cl
    d. Rb and O
    e. K and S

12. Which of the following two elements would be expected to form a covalent bond?
    a. Na, O
    b. H, S
    c. K, Te
    d. Mn, F
    e. Ba, S

13. Which of the following is a covalent compound?
    a. SCl
    b. FrCl
    c. CaS
    d. BaS
    e. KCl

14. Which of the following is an ionic compound?
    a. NO
    b. CO
    c. OF_2
    d. O_2
    e. K_2O

15. Which of the following is an ionic compound?
    a. H_2O
    b. CS_2
    c. $CaCl_2$
    d. CO_2
    e. CH_4

16. Which of the following is a covalent compound?
    a. NaCl
    b. $FeCl_3$
    c. $BaCl_2$
    d. NCl_3
    e. $AlCl_3$

17. What do you call the following compound: Na_3PO_4?
    a. a molecule
    b. a polyatomic ion
    c. a particle
    d. a molecule or a particle

IONS: IDENTIFYING CATIONS AND ANIONS

1. What is the expected charge on the Ba ion?

2. What is the expected charge on the alkaline earth metals?

3. Which of the following forms a +3 ion?
   a. Na
   b. Mg
   c. Al
   d. Si
   e. P

4. Complete the following table:

Element	Ion(s)	Oxidation Number(s)	Names
Iron	Fe^{2+} Fe^{3+}	+2 +3	iron (II) or ferrous iron (III) or ferric
Potassium			
Beryllium			
Tin			
Cobalt			
Oxygen			
Fluorine			
Copper			

5. Give the ions present and their relative numbers in K_2SO_4.
   a. 1 K^+ and 1 SO_4^-
   b. 2 K^+ and 1 SO_4^{2-}
   c. 1 K^+ and 2 SO_4^{2-}
   d. 2 K^+ and 2 SO_4^{2-}
   e. 2 K^{2+} and 1 SO_4^{2-}

6. What is the symbol for the chlorite ion?
   a. ClO^-
   b. CrO_4^-
   c. ClO_2^-
   d. ClO_3^-
   e. ClO_4^-

7. What is the symbol for the sulfite ion?
   a. SO_3^-
   b. SO_3^{2-}
   c. SO_4^{2-}
   d. SO_2^{2-}
   e. SO_4^-

8. What is the symbol for the chlorate ion?
   a. CrO_4^-
   b. ClO_2^{3-}
   c. ClO^-
   d. ClO_3^-
   e. ClO_4^-

NOMENCLATURE: WRITING FORMULAS & NAMING CHEMICAL COMPOUNDS

IONIC

1. Given the following ions, write the correct formula: Ba^{2+} and PO_4^{3-}
   a. $BaPO_4$
   b. $Ba_2(PO_4)_3$
   c. Ba_2PO_{12}
   d. Ba_3PO_8
   e. $Ba_3(PO_4)_2$

2. What is the chemical formula for lead (IV) sulfide?
   a. PbS_4
   b. PbS
   c. Pb_4S
   d. PbS_2
   e. Pb_2S

3. What is the chemical name of $Fe(HSO_3)_3$?
   a. iron (III) sulfite
   b. iron (III) bisulfite
   c. iron (I) bisulfite
   d. iron (III) sulfate
   e. iron (III) bisulfate

4. What is the chemical name of $BaSO_4$
   a. Barium sulfur
   b. Barium sulfoxide
   c. Barium sulfide
   d. Barium sulfite
   e. Barium sulfate

5. What is the chemical name of $Mg_3(PO_4)_2$?
   a. Magnesium phosphide
   b. Magnesium phosphotate
   c. Magnesium phosphate
   d. Magnesium (II) phosphate

6. Write the chemical formula for each of the following compounds:

potassium sulfate
sodium oxide
aluminum nitride
iron(II) perchlorate
calcium carbonate
chromium (III) chloride
cesium bromide
barium sulfate
beryllium oxide
sodium bicarbonate
tin (II) fluoride
cobalt (III) nitrate
potassium chlorate
lithium hydroxide
copper (I) sulfide
calcium iodide

7. Write the chemical name for each of the following compounds

Cu_2S
SnO_2
CrO_3
Cr_2O_3
Al_2O_3
$NaCl$
$MgCl_2$
$RbBr$
CsF
AlI_3
$FePO_4$
$NaHSO_4$
KNO_3
$PtCl_4$
NH_4C

NOMENCLATURE: WRITING FORMULAS & NAMING CHEMICAL COMPOUNDS MOLECULAR (COVALENT)

1. What is the correct name for N_2S?
   a. nitrogen monosulfide
   b. mononitrogen disulfide
   c. nitrogen sulfur
   d. dinitrogen disulfide
   e. dinitrogen monosulfide

2. The formula of the compound trichlorine tetrafluoride is _____.
   a. ClF
   b. Cl_3F_4
   c. Cl_3F_5
   d. Cl_5F_3
   e. Cl_4F_3

3. Write the chemical formula for each of the following compounds:

diphosphorus trioxide
arsenic pentachloride
dihydrogen monoxide
chlorine monoxide
sulfur difluoride
oxygen difluoride
sulfur dioxide

4. Write the chemical name for each of the following compounds

H_2	H_2O
CO_2	NH_3
SO_3	CH_4
P_2S_6	N_2O_4
C_2F_2	SeO_2
SeO_3	NI_3
PCl_3	SF_2
NO	NF_3
N_2F_4	H_2SO_4
SiF_4	
NH_4Cl	

NOMENCLATURE: WRITING FORMULAS & NAMING CHEMICAL COMPOUNDS ACIDS

1. What is the correct name for HCl?
   a. hydrogen sulfur
   b. chloric acid
   c. chlorous acid
   d. hydrochloric acid
   e. hydrogen chlorine

2. What is the name of H_2S?
   a. hydrogen sulfate
   b. sulfuric acid
   c. sulfurous acid
   d. hydrosulfuric acid

3. Write the chemical formula for each of the following compounds:
   a. carbonic acid
   b. hydrobromic acid
   c. hypochlorous acid
   d. hydrochloric acid
   e. nitrous acid

4. Write the chemical name for each of the following compounds:
   a. HNO_2
   b. $HBrO_2$
   c. HF
   d. HNO_3
   e. H_2SO_4

NOMENCLATURE: WRITING FORMULAS & NAMING CHEMICAL COMPOUNDS IONIC AND MOLECULAR (COVALENT) AND ACIDS

1. Write the chemical formulas for the following compounds:
   a. sodium acetate
   b. ferric bicarbonate
   c. zinc sulfite
   d. silver bicarbonate
   e. potassium iodide
   f. barium bisulfate
   g. lead(IV) chlorite
   h. nitric acid
   i. calcium sulfide
   j. lead(II) nitrite
   k. copper(I) bisulfate
   l. potassium dichromate
   m. sulfuric acid
   n. boron monophosphide
   o. cobaltic chlorate

2. Write the chemical names for the following compounds:
   a. SO_3
   b. $Be(ClO_4)_2$
   c. $(NH_4)_2Cr_2O_7$
   d. $Ba(BrO_3)_2$
   e. XeF_2
   f. Al_2S_3
   g. Na_2HPO_4
   h. $Mg_3(PO_4)_2$
   i. $Al(OH)_3$
   j. $CuSO_3$
   k. Li_2HPO_4
   l. $Ca(NO_3)_2$
   m. $Cr_2(SO_3)_3$
   n. $Ni(ClO_4)_2$
   o. $HClO$

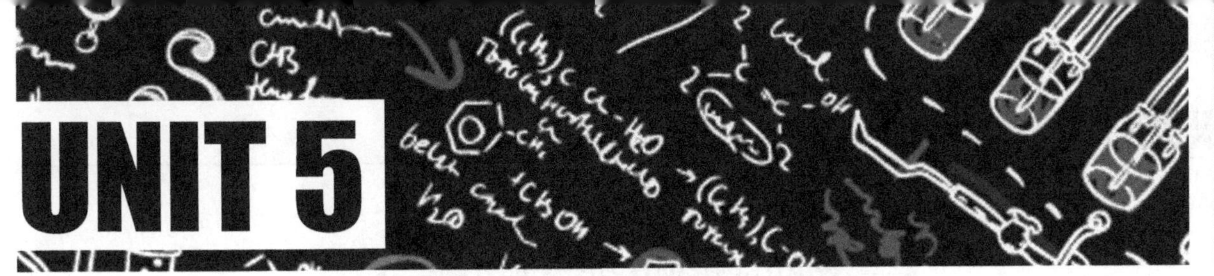

SHAPES OF COMPOUNDS (LEWIS STRUCTURES AND MOLECULAR GEOMETRY)

Valence Electrons

Lewis Dot Symbols

Lewis Dot Structures of Ionic Bonds

Lewis Dot Structures of Molecules

Molecular Geometry

- VSEPR: Valence-Shell Electron Pair Repulsion Theory
  - Linear Electron Pair (Domain) Geometry
  - Trigonal Planar Electron Pair (Domain) Geometry
  - Tetrahedral Electron Pair (Domain) Geometry
  - Trigonal Bipyramidal Electron Pair (Domain) Geometry
  - Octahedral Electron Pair (Domain) Geometry
- Molecular Polarity (Revisited)

OBJECTIVES FOR UNIT 5

- To know the fundamentals of the Lewis theory of bonding.
- To be able to write the Lewis symbols for elements.
- To give the Lewis symbols of ionic compounds.
- To write Lewis structures for covalent compounds and polyatomic ions, including those that do not follow the octet rule and those that have odd numbers of electrons.
- To know what electronegativity means and how it is used to determine the polarity of a bond.
- To understand bond lengths, bond order, and bond energies.
- To be able to use VSEPR theory to determine electron group geometry and molecular geometry.
- To be able to determine if bonds are polar and to use the molecular shape to predict whether a molecule is polar.

VALENCE ELECTRONS

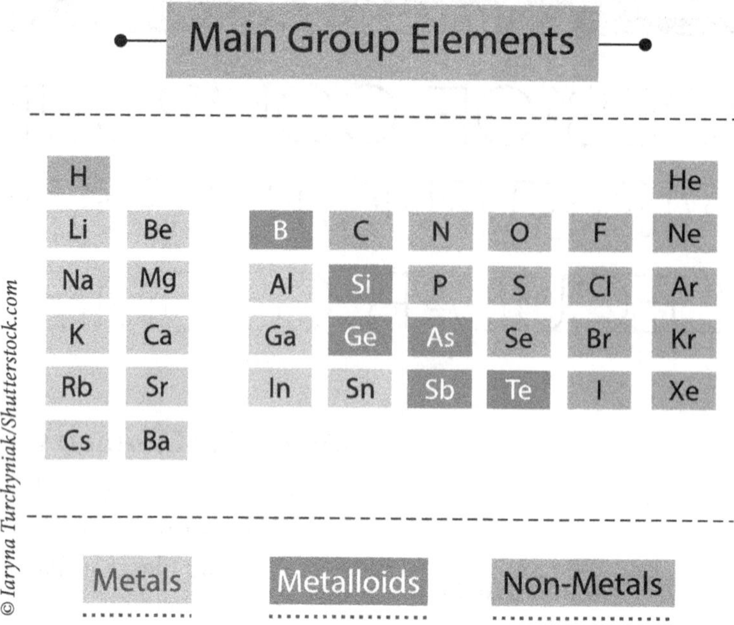

- Each column is called a group. Each element in a group has the same number of electrons in their outer orbital, also known as shells.
- The electrons in the outer shell are called valence electrons.

LEWIS DOT SYMBOLS

Lewis dot symbols describe the valence electrons present for each atom (or ion).

For representative elements:

Include the electrons in the outer most energy level

IA	IIA	IIIA	IVA	VA	VIA	VIIA	VIIIA
H·							He:
Li·	Be:	B:	C	:N:	:O	:F.	:Ne:
Na·	Mg:	Al:	Si	:P:	:S	:Cl.	:Ar:
K·	Ca:		Ge	:As:	:Se	:Br:	:Kr:
Rb·	Sr:			:Sb:	:Te	:I.	:Xe:
Cs·	Ba:				:Po	:At·	:Rn:
Fr·	Ra:						

LEWIS DOT STRUCTURES OF IONIC BONDS

Ionic bonds are formed by the transfer of one or more valence electrons from one atom to another. The ions that are formed usually have a noble gas configuration because they are stable. Generally, bonding between a metal and a nonmetal is ionic, producing high melting point solids. We call these compounds particles, salts, or formula units.

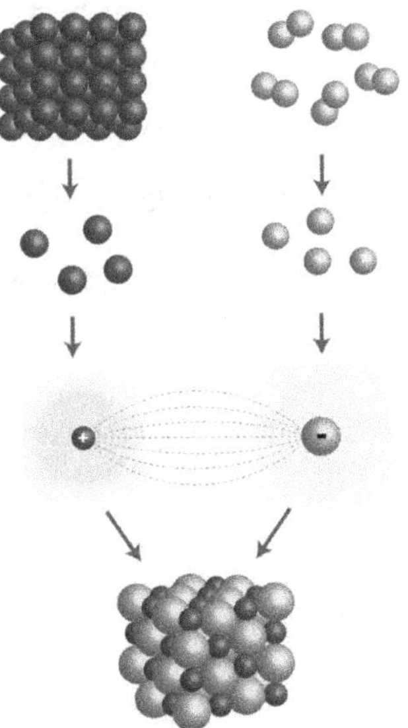

© magnetix/Shutterstock.com

Example: Barium nitride

Before Bond Ba_3N_2		After Bond: Ba_3N_2	
Ba:	N:	Ba^{2+}	$[:N:]^{3-}$
Ba:		Ba^{2+}	
Ba:	N:	Ba^{2+}	$[:N:]^{3-}$

Another way to represent after the bond:

$3Ba^{2+}$ $2[:N:]^{3-}$

A 3-D array of positive metal ions with negative nonmetal ions is called a crystal lattice.

LEWIS DOT STRUCTURES OF MOLECULES

To determine Lewis dot structures:

1. Determine the total valence electrons to be included on the structure.
2. Determine your center atom and flood remaining atoms around the center atom.

   The atomic arrangement for a molecule is usually given as follows:

   In general when there is a single central atom in the molecule, CH_2ClF, $SeCl_2$, O_3 (CO_2, NH_3, PO_4^{3-}), the central atom is the first atom in the chemical formula.

   An exception occurs when the first atom in the chemical formula is hydrogen (H) or fluorine (F), in which case the central atom is the second atom in the chemical formula.
3. Draw a bond between each pair of bonded atoms (the skeleton structure). Start off with single bond (represented by a single line showing 2 electrons shared between each atoms).
4. Use the remaining valence electrons in attempt to satisfy the octet rule.

   Follow the octet (There are exceptions): When atoms bond, most try to form an octet, which is to have 8 valence electrons around the atom (achieving a noble gas configuration). This is known as the octet rule.

   - H will only accept 2 electrons
   - B will only accept 6 electrons
   - Extra electrons place onto the center atom

   Single bond is 2 electrons shared –

   Double bond is 4 electrons shared =

   Triple bond is 6 electrons shared ≡
5. Include the charge on the Lewis dot structure (if necessary).
6. Check the arrangement with formal charges.

   Bonding pair: The electrons shared by two atoms. (This can be single, double, or triple.)

   Nonbonding pair (lone pair): The electrons not shared between atoms.

   Generally, the best structure is one that:

   - has formal charge between +1 and -1 for atoms; and
   - has no two adjacently bonded atoms with the same sign of charge.

   F.C. = (valence electrons) – (1/2 bonding electrons) – (lone electrons)

Example of writing Lewis structures:

CH_2ClF

Total valence electrons = 20e-

```
  :C̈l:
    |
H–C–F̤̈:
    |
    H
```

NO_3^-

Total valence electrons = 24e-

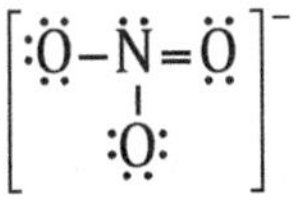

```
[:Ö–N̈=Ö ]⁻
    |
   :Ö:
```

CO_2

Total valence electrons = 16e-

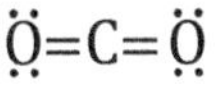

```
Ö=C=Ö
```

PO_4^{3-}

Total valence electrons = 32e-

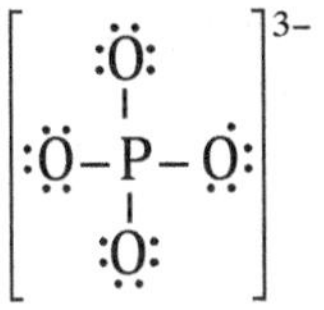

```
[    :Ö:     ]3–
      |
 :Ö – P – Ö:
      |
     :Ö:
```

Isoelectronic molecules are molecules that contain the same number and distribution of valence electrons.

Example: CO_2 and N_2O

Resonance structures exist when a compound forms more than one possible Lewis structure differing in the arrangement of electrons (not arrangement of atoms).

Example:

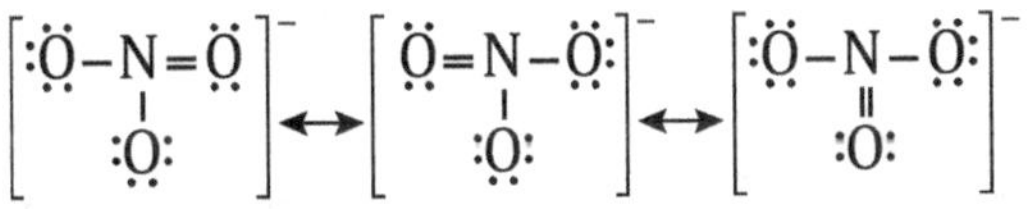

```
[:Ö–N=Ö]⁻      [Ö=N–Ö:]⁻      [:Ö–N–Ö:]⁻
    |      ↔       |      ↔       ||
   :Ö:            :Ö:            :O:
```

For compounds that do not follow the octet rule:

1. Compounds with atoms less than eight valence electrons:

Example:

```
H–B–H
  |
  H
```

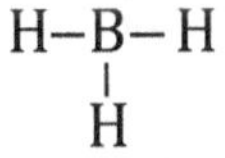

2. Compounds with atoms more than eight electrons:

Example:

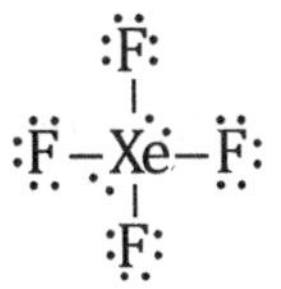

```
     :F:
      |
 :F – Xe – F:
      |
     :F:
```

Looking at the bond:

1. Bond order is the number of bonding and electron pairs shared between two atoms in a molecule. Note that as the bond order increases, the bond strength increases.
2. Bond length is the distance between the nuclei of two bonded atoms.
3. Bond energy: bond dissociation energy is the amount of energy required to separate atoms. All atoms are in gas phase when energy is measured.
4. **Note:** Shorter bonds have higher energies:

Bond	# of Electrons	Bond Order	Bond Strength	Bond Length	Bond Energy
single	2	1	weakest	longest	lowest
double	4	2			
triple	6	3	strongest	shortest	highest

Bond polarity electronegativity is a measure of the ability of an atom in a molecule to draw bonding electrons to itself. (The tendency for an atom to attract a pair of electrons in a covalent bond is called electronegativity.)

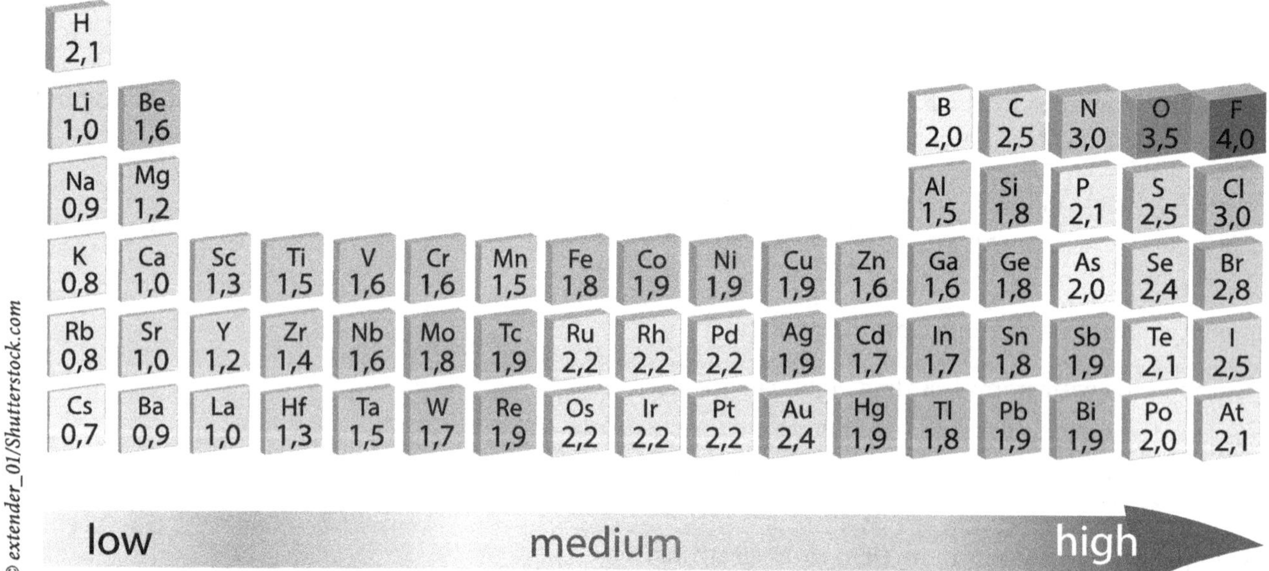

Polar molecular bond occurs when there is an unequal sharing of electrons between two atoms whose electronegativity difference is greater than 0.4.

Nonpolar molecular bond occurs when there is an equal sharing of electrons between two atoms whose electronegativity difference is less than 0.4.

Difference in electronegativity

4.0	1.7 molecular	0.4 molecular	0
Ionic bond	Polar-molecular bond	Non-polar molecular bond	
100%	50%	5%	0%

Example:

Bond between H-O

H electronegativity = 2.1

O electronegativity = 3.5

Oxygen is a more negative atom in the bond. The electronegativity difference is 1.4 making the bond polar molecular.

Note: The more polar the bond, the stronger the bond.

Dipole moment is the quantitative measure of the degree of charge separation in a molecule. This effects the polarity of a bond. Any molecule that has a net separation of charge has a dipole moment.

Example:

H-O
↦

MOLECULAR GEOMETRY

- The shape of a molecule plays an important role in its reactivity.
- By noting the number of bonding and nonbonding electron pairs on the central atom we can easily predict the shape of the molecule

What determines the shape of a molecule?

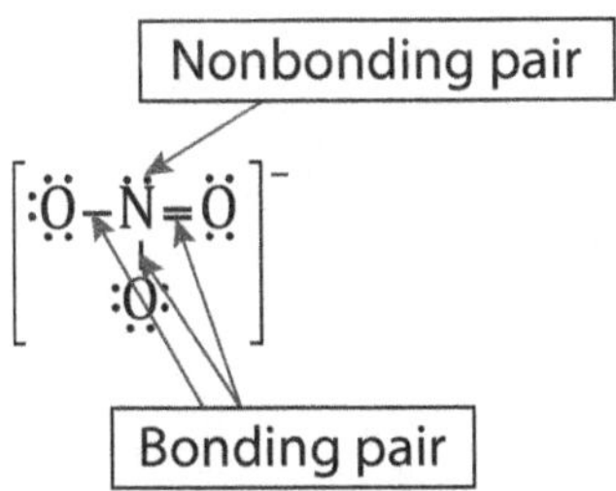

- Simply put, electron pairs, whether they be bonding or nonbonding, repel each other.
- By assuming the electron pairs are placed as far as possible from each other, we can predict the shape of the molecule.

VSEPR: VALENCE-SHELL ELECTRON PAIR REPULSION THEORY

VSEPR accounts for the geometrical arrangement of shared and unshared . s around the central atom in terms of repulsion between electron pairs (both bonding and nonbonding pairs).

The basic theory states that the best shape formed by a molecule is one that minimizes repulsion (the . s in valence shell are as far apart from each other as possible).

The shape of the molecule is described by the position of the atoms.

VSEPR treats double bonds and triple bonds as single bonds.

Guideline for applying VSEPR:

1. Write Lewis structure for molecule (or ion). To help determine the amount of bonds (both shared and unshared) around the central atom:
2. Count the number of bonding and nonbonding pairs around the central atom.

Electron Domains

This molecule has four electron domains.

- We can refer to the electron pairs as electron domains.
- In a double or triple bond, all electrons shared between those two atoms are on the same side of the central atom; therefore, they count as one electron domain.

Example: NH_3

$NH_3 \rightarrow$ H–$\ddot{N}$–H (with H below N) $\rightarrow$ H–N(–H)(–H) $\longrightarrow$

© extender_01/ Shutterstock.com

NH_3 is tetrahedral in the electron pair geometry and trigonal pyramidal in the molecular geometry

The following are the possible electron-pair geometries:

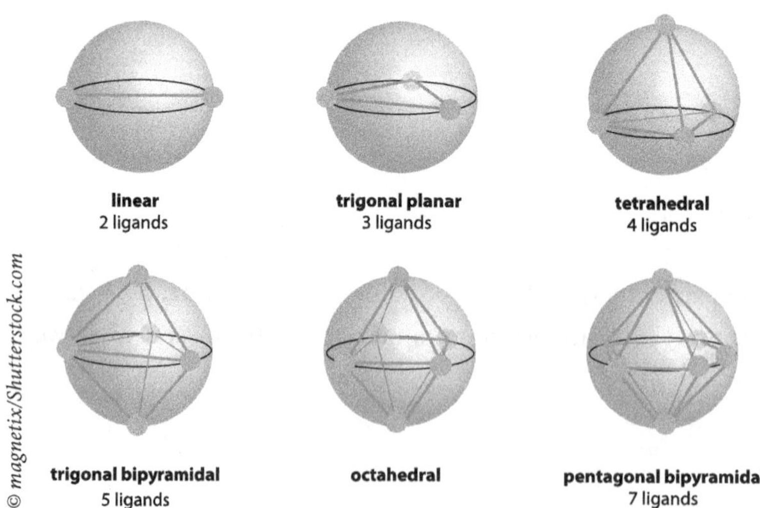

Linear, Trigonal Planar, Tetrahedral, Trigongal Bipyramidal, Octahedral, and pentagonal bipyramidal

LINEAR ELECTRON-PAIR (DOMAIN) GEOMETRY

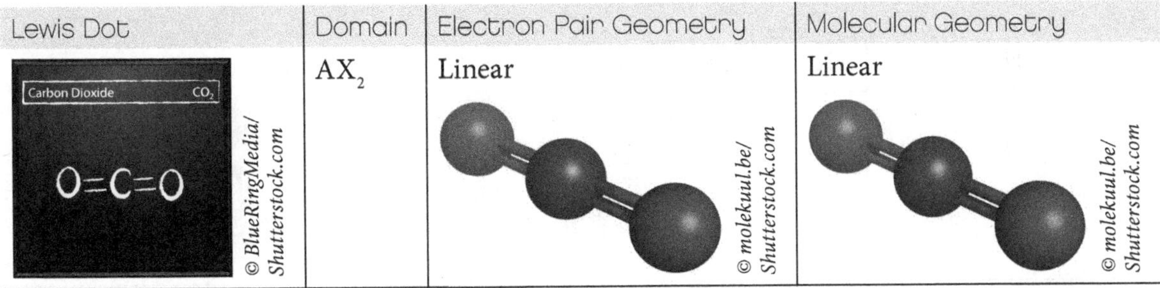

Lewis Dot	Domain	Electron Pair Geometry	Molecular Geometry
Carbon Dioxide CO_2 O=C=O © BlueRingMedia/ Shutterstock.com	AX_2	Linear © molekuul.be/ Shutterstock.com	Linear © molekuul.be/ Shutterstock.com

- In this domain, there is only one molecular geometry: linear.
- Because there are only two attachments to the center atom, the molecule will be linear no matter what the electron domain is.

TRIGONAL PLANAR ELECTRON PAIR (DOMAIN) GEOMETRY

Lewis Dot	Domain	Electron Pair Geometry	Molecular Geometry
H–B–H with H below B	AX_3	Trigonal Planar 120° © Pratchaya Ruenyen/ Shutterstock.com	Trigonal Planar H B H H © foxterrier2005/ Shutterstock.com
$[:\ddot{O}-\dot{N}=\ddot{O}]^-$	AX_2E	Trigonal Planar 120° © Pratchaya Ruenyen/ Shutterstock.com	Bent © molekuul.be/ Shutterstock.com

There are two molecular geometries:

- Trigonal planar, if all the electron domains are bonding
- Bent, if one of the domains is a nonbonding pair.

TETRAHEDRAL ELECTRON PAIR (DOMAIN) GEOMETRY

Lewis Dot	Domain	Electron Pair Geometry	Molecular Geometry
H | H–C–H | H	AX_4	Tetrahedral 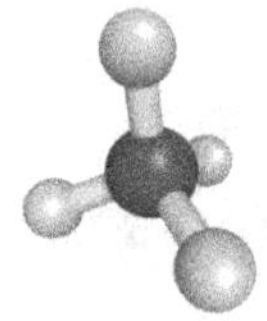 © Iculig/ Shutterstock.com	Tetrahedral 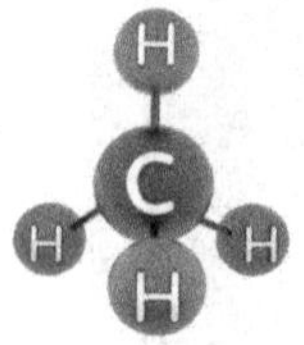 © Pankratov Yuriy/ Shutterstock.com
H–N̈–H | H	AX_3E	Tetrahedral 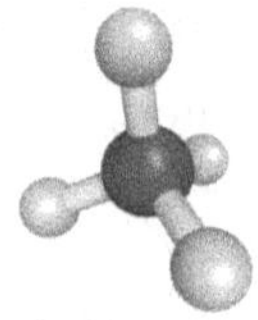 © Iculig/ Shutterstock.com	Trigonal Pyramidal 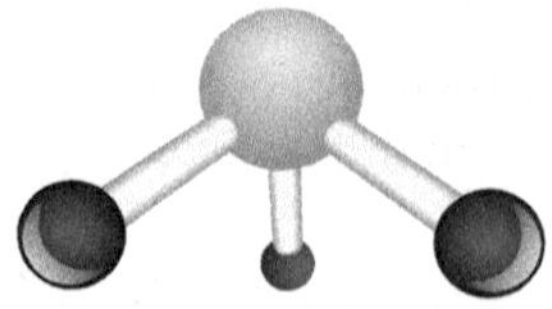 © extender_01/ Shutterstock.com
H–Ö–H (with two lone pairs on O)	AX_2E_2	Tetrahedral 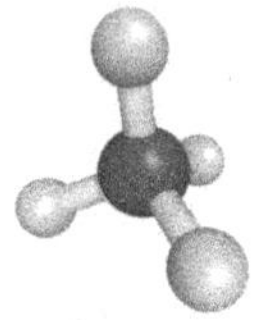 © Iculig/ Shutterstock.com	Bent  © foxterrier2005/ Shutterstock.com

There are three molecular geometries:

- Tetrahedral if all are bonding pairs
- Trigonal pyramidal if one is a nonbonding pair
- Bent if there are two nonbonding pairs

TRIGONAL BIPYRAMIDAL ELECTRON PAIR (DOMAIN) GEOMETRY

Lewis Dot	Domain	Electron Pair Geometry	Molecular Geometry
:Cl: :Cl: :Cl–P–Cl: :Cl:	AX_5	Trigonal Bipyramidal	Trigonal Bipyramidal
:F: :F–S–F: :F:	AX_4E	Trigonal Bipyramidal	Seesaw
:F–Cl–F: :F:	AX_3E_2	Trigonal Bipyramidal	T-Shaped
:F–Xe–F:	AX_2E_3	Trigonal Bipyramidal	Linear

There are four distinct molecular geometries in this domain:

- Trigonal bipyramidal
- Seesaw
- T-shaped
- Linear

OCTAHEDRAL ELECTRON PAIR (DOMAIN) GEOMETRY

Lewis Dot	Domain	Electron Pair Geometry	Molecular Geometry
:F: :F: :F–S–F: :F: :F:	AX_6	Octahedral	Octahedral
:F: :F: :F–Br–F: :F:	AX_5E	Octahedral	Square Pyramidal
:F: :F–Xe–F: :F:	AX_4E_2	Octahedral	Square Planar

All positions are equivalent in the octahedral domain.

There are three molecular geometries:

- Octahedral
- Square pyramidal
- Square planar

MOLECULAR POLARITY (REVISITED)

If two bond dipoles of equal magnitude (numerical value is the same), but opposite direction cancel each other out, you have the possibility of having a nonpolar molecule.

A molecule is nonpolar if:

1. All the X atoms (or groups) are arranged symmetrically around the central atom a in the molecular geometries: linear, trigonal-planar, tetrahedral, trigonal bypyramidal, octahedral
2. All the X atoms (or groups) are the same.
3. All the X atoms (or groups) have the same partial charges.

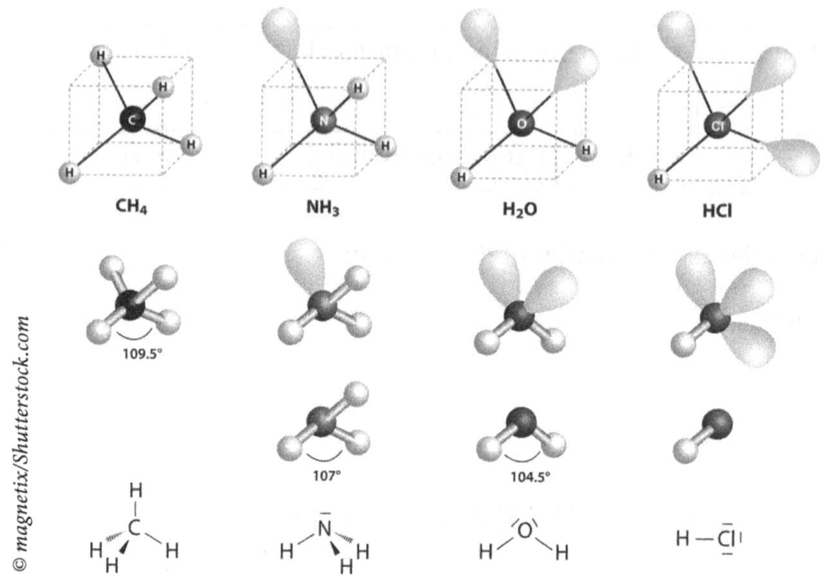

IONIC BONDING LEWIS DOT STRUCTURES

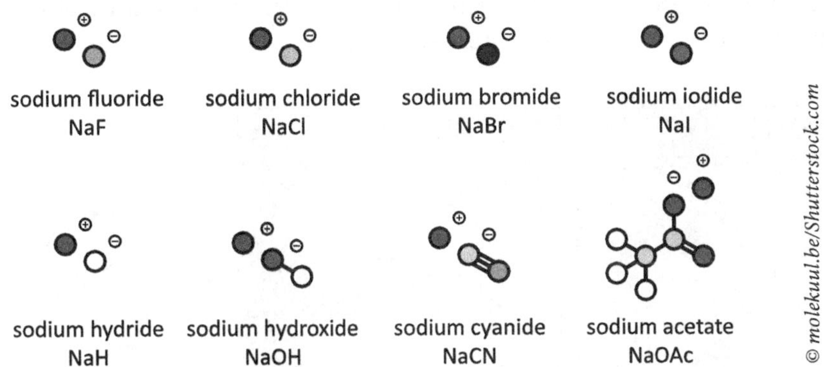

Ions are formed where electrons are transferred from the valence shell of one atom (usually a metal) to the valence shell of another atom (non-metal) so that both end up with noble gas configurations. Assume, in the first instance, that compounds between reactive metals and reactive non-metals will be ionic.

1. Draw diagrams (outer electrons only) to show the bonding in the following ionic compounds. Draw a before and after bond picture for each on another sheet of paper.
   a. Lithium fluoride—LiF

   b. Magnesium sulphide—MgS

   c. Calcium chloride—$CaCl_2$

   d. Sodium oxide—Na_2O

   e. Aluminium oxide—Al_2O_3

   f. Magnesium nitride—Mg_3N_2

MOLECULAR BONDING LEWIS DOT STRUCTURE

Molecular bonding involves the sharing of electron pairs between two atoms. This occurs most often between non-metal atoms, but there are a number of compounds between metals and non-metals that are molecular. A single molecular bond involves one shared pair of electrons. In many compounds, atoms will share electrons to enable their valence shell to become like the nearest noble gas. This is normally 8 electrons (the Octet Rule), apart from hydrogen. There are exceptions (see next section).

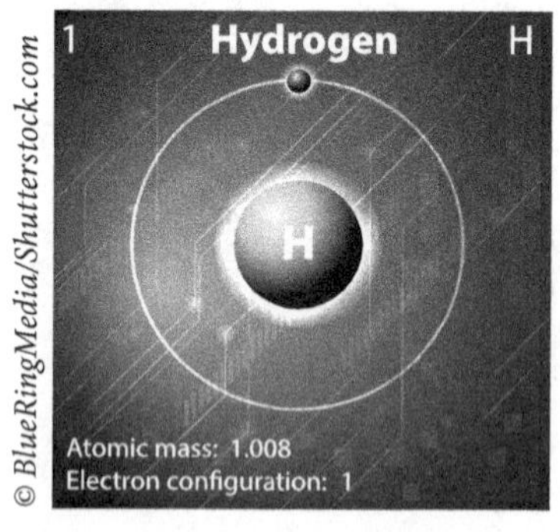

© BlueRingMedia/Shutterstock.com

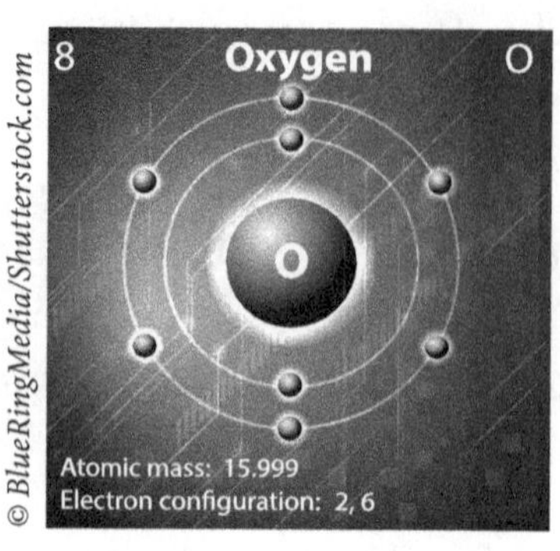

© BlueRingMedia/Shutterstock.com

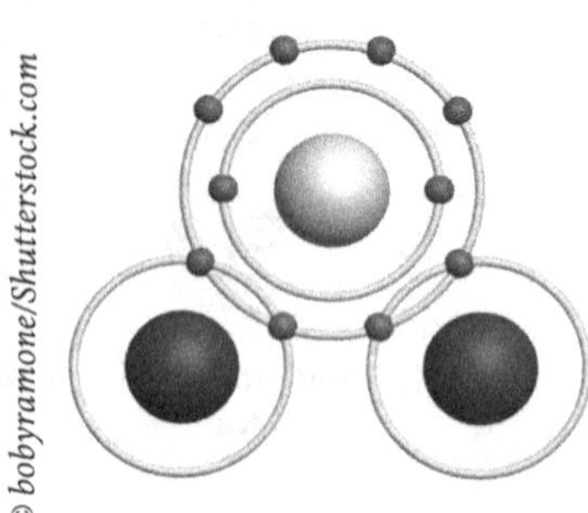
© bobyramone/Shutterstock.com

1. Draw diagrams (outer electrons only) to show the bonding in the following molecules. Draw a before and after bond picture for each on another sheet of paper. (Your drawings should be similar to the example above.)
   a. Hydrogen fluoride—HF

   b. Chlorine—Cl_2

   c. Oxygen—O_2

   d. Nitrogen—N_2

   e. Cyanide—CN^-

MOLECULAR GEOMETRY

Complete the Following Table:

Compound	Lewis Dot	Domain	Electron Pair	Molecular
CO_2				
BF_3				
NO_2^-				
CH_4				
NH_3				
H_2O				
PCl_5				
SF_4				

Compound	Lewis Dot	Domain	Electron Pair	Molecular
ClF_3				
XeF_2				
SF_6				
BrF_5				
XeF_4				

CHEMISTRY DISCOVER UNIT 8
LEWIS STRUCTURES AND MOLECULAR GEOMETRY

Compound	Lewis Dot	Domain	Electron Pair	Molecular
SO_2				
BCl_3				
NH_3				

Compound	Lewis Dot	Domain	Electron Pair	Molecular
PO_4^{3-}				
NO_3^-				
SO_3				
SO_3^{2-}				
PF_3				
SF_6				
BrF_5				
BrO_3^-				
ICl_4				
XeF_2				

ELECTRONEGATIVITY

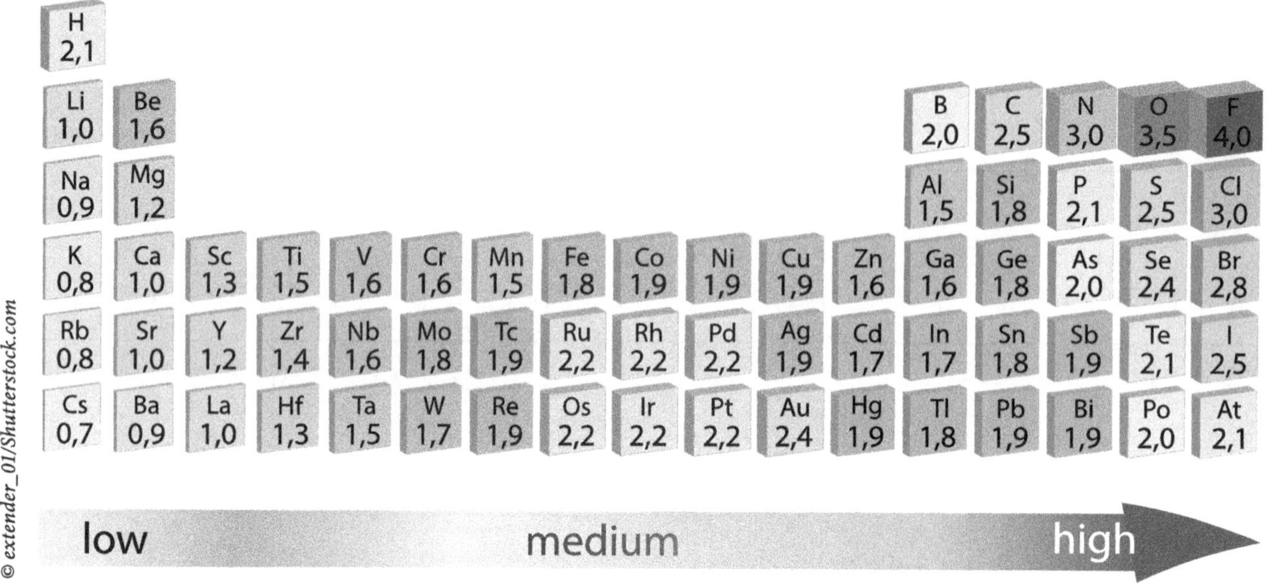

4.0	1.7 molecular	0.4 molecular	0
Ionic bond	Polar-molecular bond	Non-polar molecular bond	
100%	50%	5%	0%

Bonding Between	More Electronegative Element and Value	Less Electronegative Element and Value	Difference in Electronegativity	Bond Type
Sulfur and hydrogen				
Sulfur and cesium				
Chlorine and bromine				
Calcium and chlorine				
Oxygen and hydrogen				
Nitrogen and hydrogen				
Iodine and iodine				
Copper and sulfur				
Hydrogen and fluorine				
Carbon and oxygen				

LEWIS DOT STRUCTURES AND MOLECULAR GEOMETRY

1. How many valence electrons does gallium have?
   a. 1
   b. 3
   c. 5
   d. 13
2. How many electrons does phosphorus have to gain in order to achieve a noble gas electron configuration?
   a. 2
   b. 3
   c. 5
   d. 4
3. How many valence electrons are transferred from the nitrogen atom to each potassium in the formation of the compound potassium nitride?
   a. 0
   b. 1
   c. 3
   d. 5

4. Which of the following takes place in an ionic bond?
   a. Two atoms share two electrons
   b. Two atoms share electrons such that both follow the octet rule.
   c. Like-charged ions attract
   d. Oppositely-charged ions attract

5. What is the net charge of the ionic compound calcium fluoride?
   a. −2
   b. 0
   c. +2
   d. −1

6. How many electrons are shared between two atoms in a double molecular (covalent bond?)
   a. 8
   b. 6
   c. 4
   d. 2

7. How many unshared pairs of electrons are there in hydrogen iodide?
   a. 1
   b. 6
   c. 3
   d. none of these

8. Which of the following elements occurs naturally as a diatomic molecule with a triple bond?
   a. oxygen
   b. hydrogen
   c. fluorine
   d. nitrogen

9. Draw the 2-D Lewis structure below the molecular formula.
   - Determine both electron pair (EP) and molecular geometry.
   - From the overall molecular geometry and the presence and arrangement of polar bonds (if any), determine if a molecule is polar. (Polarity does not apply to polyatomic ions.)

a. PF_3	i. ClO_4^-	q. BH_3
b. PF_5	j. KrF_2	r. XeF_3^+
c. SF_4	k. XeF_4	s. BrF_3
d. SF_6	l. XeO_3	t. IF_5
e. CH_3^+	m. XeO_2F_2	u. HCN
f. ClO^-	n. CS_2	v. NH_3Cl^+
g. ClO_2^-	o. NO_3^-	w. I_3^-
h. ClO_3^-	p. CO_3^{2-}	x. N_3^-
		y. $BHCl_2$

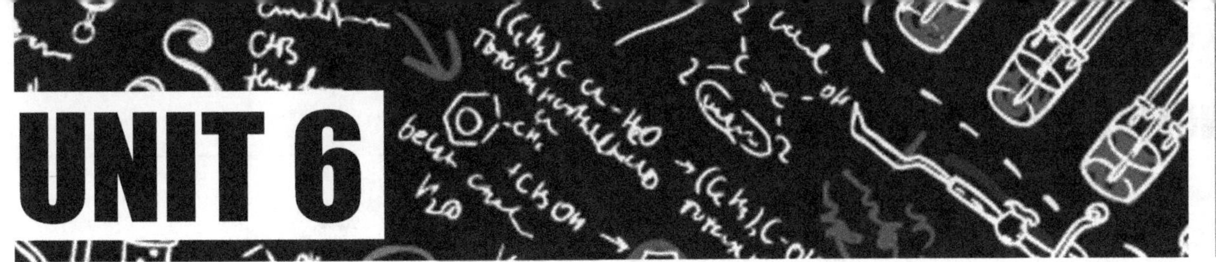

THE MIGHTY MOLE (CHEMICAL FORMULA RELATONSHIPS)

The Number of Atoms in a Formula

Atomic Mass, Molecular Mass, and Formula Mass

The Mole Concept

Molar Mass

Conversions with Mass, Number of Moles, and Number of Units

Percentage Composition (Mass Relationship Between Elements in a Compound)

Empirical Formula of a Compound

Empirical Formula Calculation
Molecular Formula Calculation

OBJECTIVES FOR UNIT 6

- To be able to name or give the formula of binary molecular compounds.
- To use the periodic table for calculating molecular and formula masses.
- To know the value and meaning of Avogadro's number.
- Given the number of moles, to be able to calculate the number of grams of a substance and vice versa, using the molar mass.
- Given the number of grams or moles of a substance, to be able to calculate the number of particles and vice versa, using the molar mass and Avogadro's number.
- Given the formula of a substance, to be able to calculate the mass percent of each element in the formula.
- To determine empirical and molecular formulas from composition data.

THE NUMBER OF ATOMS IN A FORMULA

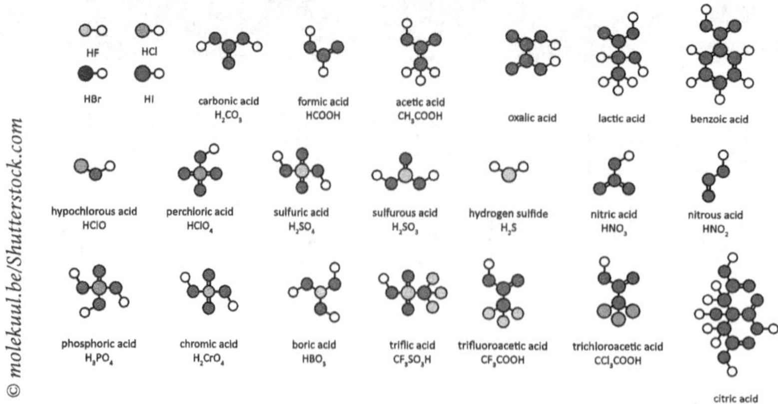

© molekuul.be/Shutterstock.com

A shorthand notation for representing chemical compounds is to write a chemical formula.

- For a single element: iron is represented by the symbol Fe indicating one atom of iron.

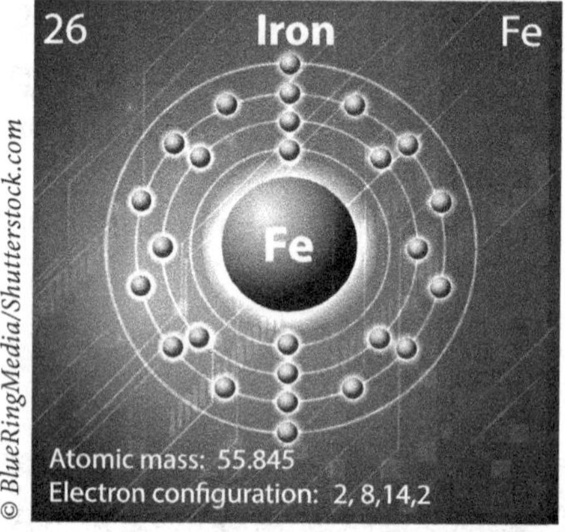

© BlueRingMedia/Shutterstock.com

- For a particle (salt or formula unit): sodium bicarbonate is represented by the formula $NaHCO_3$ indicating 1 atom sodium, 1 atom hydrogen, 1 atom carbon, and 3 atoms of oxygen.

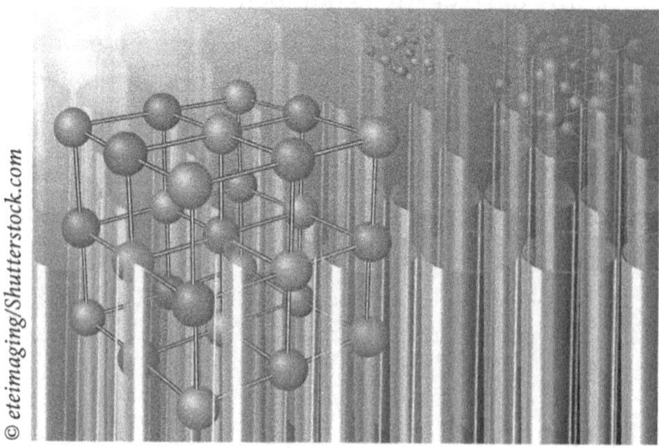
© eteimaging/Shutterstock.com

- For a molecule: sulfuric acid is represented by the formula H_2SO_4 indicating 2 atoms of hydrogen, 1 atom sulfur and 4 atoms of oxygen.

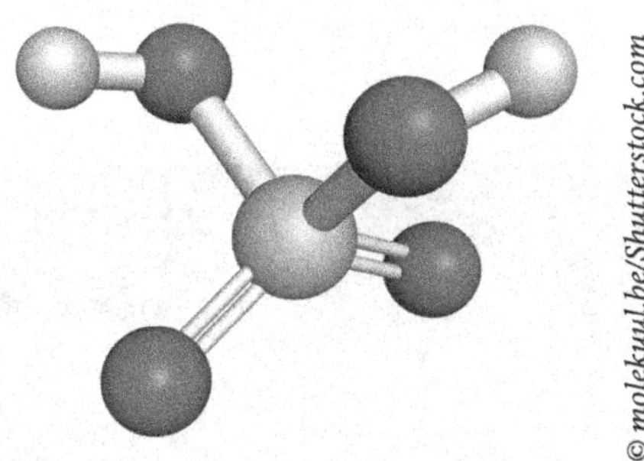
© molekuul.be/Shutterstock.com

ATOMIC MASS, MOLECULAR MASS, AND FORMULA MASS

We learned that the atomic mass (or atomic weight) that is written on the periodic table represents the average relative mass of the isotopes for each element.

For example: Molybdenum, element 42, is a silvery-white, hard, transition metal. Scheele discovered it in 1778. It was often confused with graphite and lead ore. Molybdenum is used in alloys, electrodes, and catalysts. The World War II German artillery piece called "Big Bertha" contained molybdenum as an essential component of its steel.

© Zbynek Burival/Shutterstock.com

© Sebastian Tomus/Shutterstock.com

$_{42}Mo$ has an atomic mass of 95.94 amu

When atoms combine ionically or molecularly they form particles or molecules. The atomic masses combine for these elements for determining a mass for the compound. We call this mass the formula mas (for ionic compounds) or molecular mass (for molecular compounds).

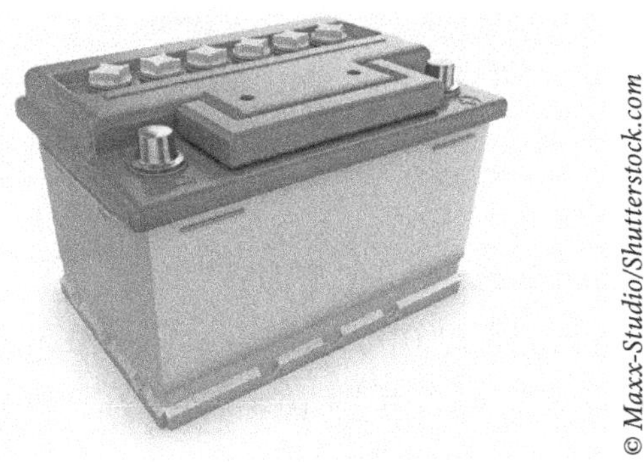
© Maxx-Studio/Shutterstock.com

Lead(II) sulfate: $PbSO_4$ is a chemical in lead-acid car batteries.

Atomic masses: Pb = 207.19 amu, S = 32.064 amu, O = 15.999 amu

Formula mass: (1 × 207.19) + (1 × 32.064) + (4 × 15.999) = 303.25 amu

For a hydrate:

Fe_2O_3 is a hematite mineral (compound for rust) which has a rhombohedral crystal (a hexagonal) system. When it combines with water, it can form the hydrate: $Fe_2O_3{\bullet}3H_2O$ which is a limonite mineral, an amorphous mineral-like substance.

Atomic masses: Fe = 55.8 amu, O = 16.0 amu, H = 1.0 amu

2 Fe = (2 × 55.8) = 111.6 amu

3 O = (3 × 16.0) = 48.0 amu

3 H_2O = (6 × 1.0) + (3 × 16.0) = 54.0 amu

Formula mass: 111.6 + 48.0 + 54.0 = 213.6 amu

Molecular mass (for molecular compounds)

Vitamin C is the molecule ascorbic acid: $HC_6H_7O_6$

Atomic masses: C = 12.011 amu, H = 1.008 amu, O = 15.999 amu

Molecular mass: (6 × 12.011) + (8 × 1.008) + (6 × 15.999) = 176.124 amu

THE MOLE CONCEPT

© VIGE.CO/Shutterstock.com

What is a mole?

To the chemistry illiterate, a mole is a furry brown animal that burrows through the ground and ruins a lot of nicely groomed lawns.

To the chemist, a mole is just a number: 6.02×10^{23} which amounts to 602,000,000,000,000,000,000,000

A little history. Back in the 19th century an Italian scientist, Amadeo Avogadro, hypothesized that equal volumes of gases, under the same conditions of temperature and pressure, contain the same number of molecules.

© rook76/Shutterstock.com

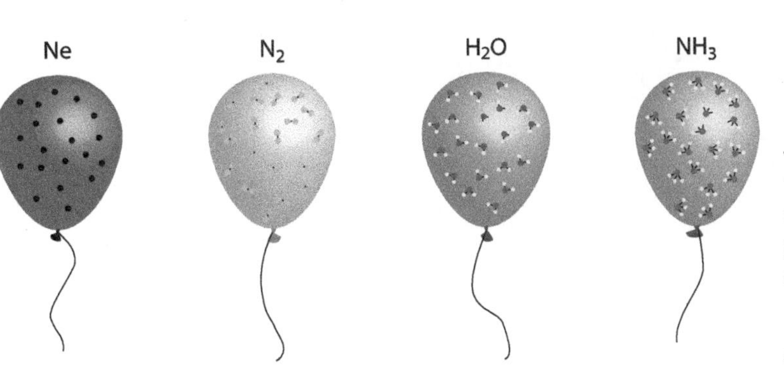

© magnetix/Shutterstock.com

Later this number of molecules was determined to be 6.02×10^{23} if the mass of the gas was equal to its atomic weight (mass) in grams. Today, chemists still rely on the quantity known as the mole, which contains Avogadro's number of formula unit.

The basic idea is that atoms, molecules, and particles are so small that it would be impossible to place one on a balance to get its mass.

Therefore, place 602,214,129,000,000,000,000,000 atoms or 602,214,129,000,000,000,000,000 molecules or 602,214,129,000,000,000,000,000 particles on a balance and get their masses.

And surprise! These masses are the same number as the atomic masses found on the periodic table.

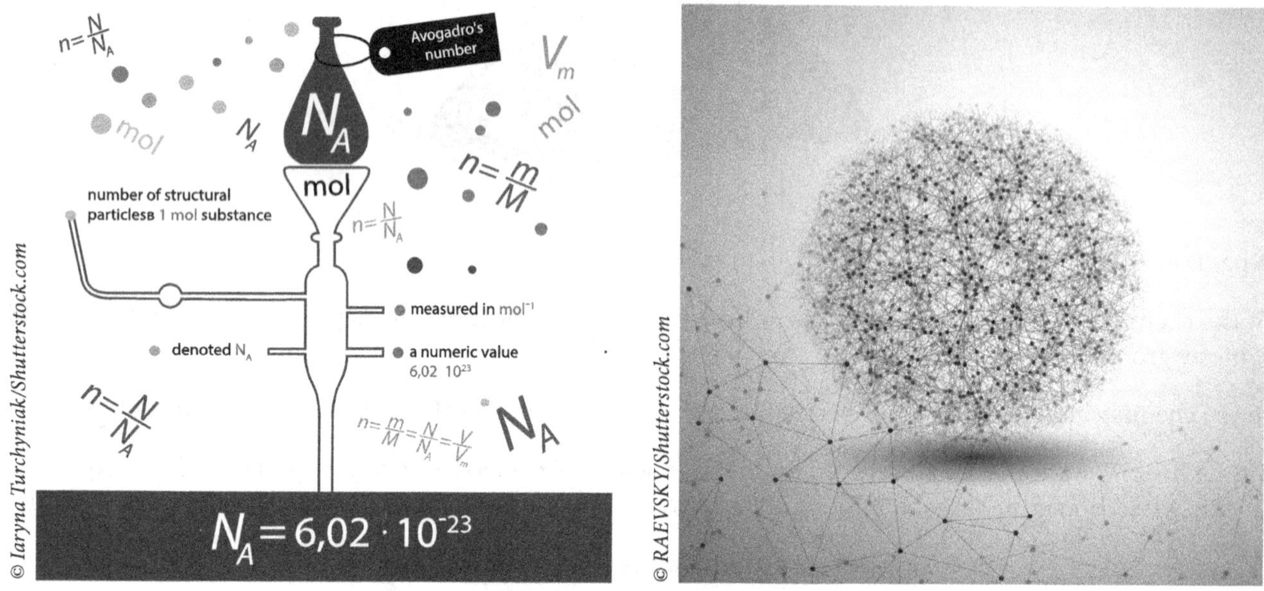

Place 6.02×10^{23} atoms of gold on a balance and you have 196.97 grams of gold.

Place 6.02×10^{23} molecules of water on a balance and you have 18.02 grams of water.

Place 6.02×10^{23} particles of NaCl (salt) and you have 58.44 grams of salt.

Now it is your turn to conjure up some moles.

Just as one would say, 1 dozen = 12, one can say, 1 mole = 6.02×10^{23}.

MOLAR MASS

How about compounds? Can we determine the mass of 6.02×10^{23} molecules or 6.02×10^{23} particles? We most certainly can!

1 mole = 6.02×10^{23} molecules = molar mass in grams

or

1 mole = 6.02×10^{23} particles = molar mass in grams

© monticello/Shutterstock.com

C_2H_5OH, ethanol, or ethyl alcohol is common alcohol for consumption.

Molar mass: 2C = (2 × 12.0) = 24.0
6H = (6 × 1.0) = 6.0
1O = (1 × 16.0) = 16.0
24.0 + 6.0 + 16.0 = 46.0 g/mol

Note: Before we wrote this as 46.0 amu. Now, thanks to the mole we can write this number as a molar mass or 46.0 g/mole.

Did you know: 1mol C_2H_5OH = 2 mol C, 6 mol H, 1 mol O

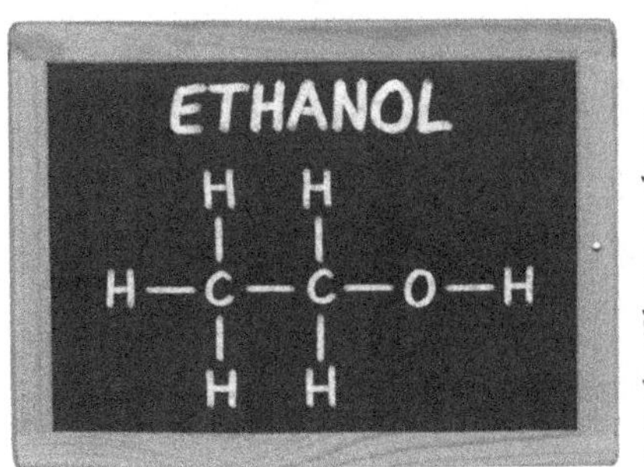

© Zerbor/Shutterstock.com

CONVERSIONS WITH MASS, NUMBER OF MOLES, AND NUMBER OF UNITS

Now that we know about the mole, we can go so much further.

We know that if we have 58.5 grams of NaCl, we have 1 mole of NaCl or 6.02×10^{23} particles of NaCl.

What if we had more or less NaCl? What if we had 33.2 grams of NaCl or 56,600 grams NaCl? How many moles of NaCl is this? How many particles of NaCl is this? Do we need to put on our special glasses and count all those particles? You can go right ahead, but I am going to use calculations to solve this problem.

Using the molar mass of substances it is possible to convert between grams (the amount in mass) of substance and moles (the amount in quantity) of a substance. Using Avogadro's number it is possible to convert between moles of a substance and atoms (or molecules or particles) of a substance.

$$\begin{bmatrix}\text{mass}\\\text{(grams)}\end{bmatrix}\underset{\times\text{molar mass}}{\overset{\div\text{molar mass}}{\rightleftarrows}}[\text{moles}]\underset{\div 6.02\times10^{23}}{\overset{\times 6.02\times10^{23}}{\rightleftarrows}}\begin{bmatrix}\text{atoms}\\\text{molecules}\\\text{particles}\end{bmatrix}$$

There are six different mole calculations to learn:

1. Given mass (grams) of substance determine moles

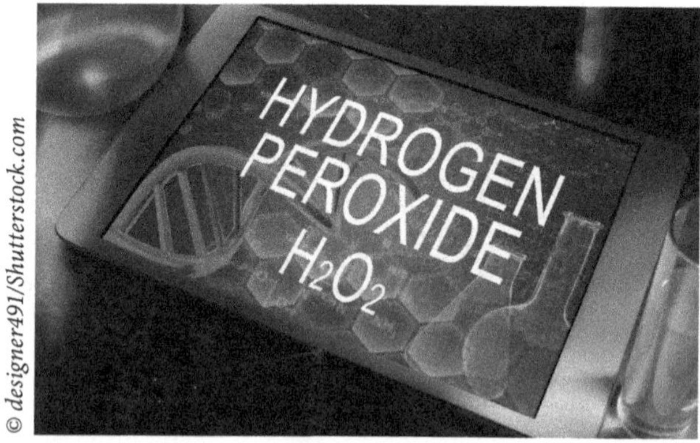

© designer491/Shutterstock.com

Example: Determine the moles in 67.89 grams of dihydrogen dioxide.

Convert 67.89 grams to moles:

What is needed is the molar mass of H_2O_2 is 34.02 g/mole (This means 34.02 g H_2O_2 = 1 mole H_2O_2)

$$67.89\text{g } H_2O_2 \times \frac{1\text{mol } H_2O_2}{34.02\text{g}} = 1.996\text{mol } H_2O_2$$

2. Given moles of substance determine mass (grams)

© EcoPimStudio/Shutterstock.com

Example: Calculate the mass of 0.28 moles caffeine $C_8H_{10}N_4O_2$

Convert 0.28 moles to grams

What is needed is the molar mass of $C_8H_{10}N_4O_2$ is 194.19 g/mole

(This means 194.19 g $C_8H_{10}N_4O_2$ = 1 mole $C_8H_{10}N_4O_2$)

$$0.28\,\text{mol}\ C_8H_{10}N_4O_2 \times \frac{194.19\,\text{g}\ C_8H_{10}N_4O_2}{1\,\text{mol}\ C_8H_{10}N_4O_2} = 54\,\text{g}\ C_8H_{10}N_4O_2$$

3. Given moles of substance determine atoms (single element), particles (Ionic), molecules (molecular)

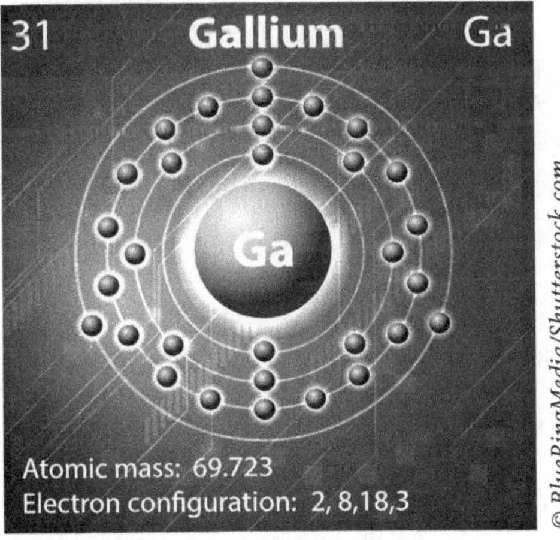

© BlueRingMedia/Shutterstock.com

Example: Calculate the quantity of atoms in 35.77 moles in gallium.

What is needed is the amount of atoms in a mole which is 6.02×10^{23}.

(This means 1 mole Ga = 6.02×10^{23} atoms)

$$35.77\,\text{mol Ga} \times \frac{6.02 \times 10^{23}\ \text{atoms Ga}}{1\,\text{mol Ga}} = 2.153 \times 10^{25}\ \text{atoms Ga}$$

4. Given atoms (single element), particles (Ionic), molecules (molecular) of substance determine moles

© bouzou/Shutterstock.com

Example: Determine the mole quantity in 5.95×10^{24} molecules carbon tetrachloride.

What is needed is the amount of molecules in a mole which is 6.02×10^{23}

(This means 1 mole $CCl_4 = 6.02 \times 10^{23}$ molecules CCl_4)

$$5.95 \times 10^{24} \text{ molecules } CCl_4 \times \frac{1 \text{mole } CCl_4}{6.02 \times 10^{23} \text{ molecules } CCl_4} = 9.88 \text{moles } CCl_4$$

5. Given mass (grams) of substance determine atoms (single element), particles (Ionic), molecules (molecular)

© OTOBOR/Shutterstock.com

Convert 6235 grams of octane (C_8H_{18}) to molecules of octane.

Step 1: Convert 6235 grams to moles:

What is needed is the molar mass of C_8H_{18} is 34.02 g/mole (This means 114 g C_8H_{18} = 1 mole C_8H_{18})

$$6235 g\ C_8H_{18} \times \frac{1 \text{mol } C_8H_{18}}{114.0 g\ C_8H_{18}} = 54.6 \text{mol } C_8H_{18}$$

Step 2: Convert 54.69 moles to molecules:

$$54.69 \text{mol } C_8H_{18} \times \frac{6.02 \times 10^{25} \text{ molecules } C_8H_{18}}{1 \text{mol } C_8H_{18}} = 3.29 \times 10^{25} \text{ molecules } C_8H_{18}$$

6. Given atoms (single element), particles (ionic), molecules (molecular) of substance determine mass (grams)

$Al_2(SO_4)_3$, aluminum sulfate, also known as alum is used in natural dying of fabrics.

How many grams are there in 9.81×10^{24} particles of aluminum sulfate?

Step 1: Convert 9.81×10^{24} particles $Al_2(SO_4)_3$ to moles $Al_2(SO_4)_3$

$$9.81 \times 10^{24} \text{ particles } Al_2(SO_4)_3 \times \frac{1\,\text{mol } Al_2(SO_4)_3}{6.02 \times 10^{25} \text{ particles } Al_2(SO_4)_3} = 16.4\,\text{mol } Al_2(SO_4)_3$$

Step 2: Convert moles $Al_2(SO_4)_3$ to grams $Al_2(SO_4)_3$

$$16.4\,\text{mol } Al_2(SO_4)_3 \times \frac{342.3\,\text{g } Al_2(SO_4)_3}{1\,\text{mol } Al_2(SO_4)_3} = 5610\,\text{g } Al_2(SO_4)_3$$

PERCENTAGE COMPOSITION (MASS RELATIONSHIP BETWEEN ELEMENTS IN A COMPOUND)

Understanding percentage composition may be helpful the next time you take a trip to the garden shop to buy fertilizers. Most fertilizers contain phosphorus, nitrogen, and potassium. The packages usually list these elements as percentages. What does this mean?

Percent composition of a compound is a calculation for determining the percent mass each element contributes to the overall mass of the compound.

The formula:

$$\% \text{ element} = \frac{\text{total mass element}}{\text{total mass of compond}} \times 100$$

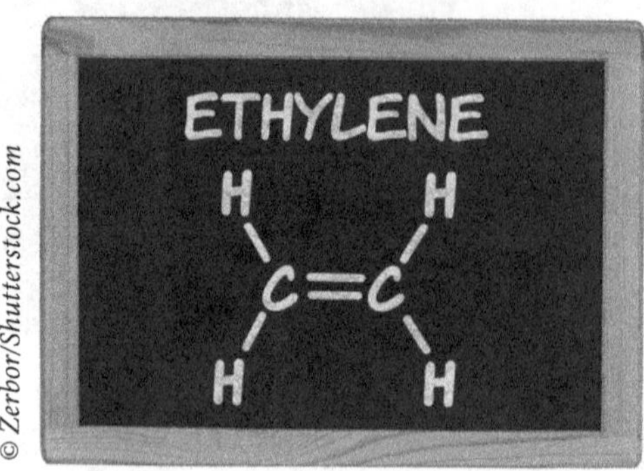

© Zerbor/Shutterstock.com

C_2H_4, ethylene, is a derivative of the plastic, polyethylene. This is a common plastic for many containers. Ethylene is produced in greater quantity than any other organic chemical in the United Staes. Calculate the mass percent of ethylene.

Mass of compound (molecular mass) = (2 × 12.0) + (4 × 1.0) = 28.0

$$\%C = \frac{24.0\text{g C}}{28.0\text{g } C_2H_4} \times 100 = 85.7\%\text{C}$$

$$\%H = \frac{4.0\text{g H}}{28.0\text{g } C_2H_4} \times 100 = 14.3\%\text{ H}$$

© Swapan Photography/Shutterstock.com

If a bag of fertilizer indicates 4.0% potash (K_2O), what percent of potassium (K) is actually in the fertilizer?

$$\%K = \frac{78.2\text{g K}}{94.0\text{g } K_2O} \times 100 = 83.0\%$$

83% of 4% = 3.32% K in bag

EMPIRICAL FORMULA OF A COMPOUND

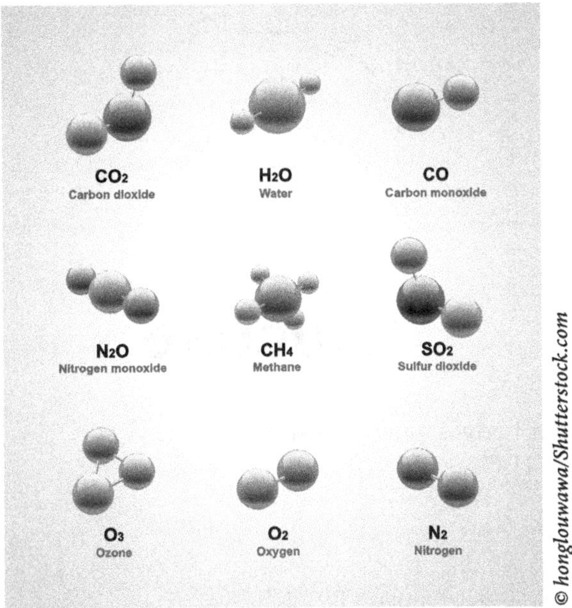

An empirical formula is the simplest ratio of atoms in a compound.

O OH

Linoleic acid

Olive oil is composed of many fatty acid components. Olive oil contains polyunsaturated acids such as $C_{18}H_{32}O_2$, Linoleic acid. Its empirical formula is written: $C_9H_{16}O$.

Note: It is different by a multiple of 2.

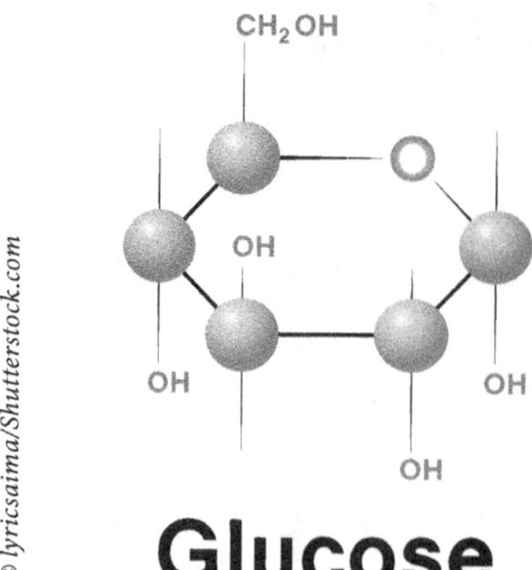

An important carbohydrate in our body is glucose, a simple sugar, $C_6H_{12}O_6$. Its empirical formula is written: CH_2O.

Note: It is different by a multiple of 6.

EMPIRICAL FORMULA CALCULATION

Nicotine, a chemical commonly found in cigarettes, contains: 74.0% C, 8.70% H, and 17.3% N.

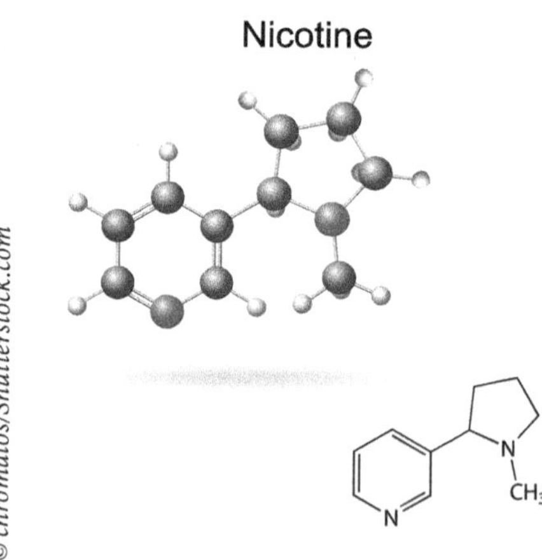

To calculate the empirical formula for nicotine:

Step 1: Convert percent to grams: 74.0 g C, 8.70 g H, 17.3 g N

Step 2: Convert grams to moles

$$74.0\text{g C} \times \frac{1\text{mol C}}{12.0\text{g C}} = 6.17\text{mol C}$$

$$8.70\text{g H} \times \frac{1\text{mol H}}{1.0\text{g H}} = 8.70\text{mol H}$$

$$17.3\text{g N} \times \frac{1\text{mol N}}{14.0\text{g N}} = 1.24\text{mol N}$$

Step 3: Determine simplest whole number ratio

$$\frac{6.17\text{mol C}}{1.24} = 4.98 \approx 5\text{mol C}$$

$$\frac{8.70\text{mol H}}{1.24} = 7.02 \approx 7\text{mol H}$$

$$\frac{1.24\text{mol N}}{1.24} = 1.00 = 1\text{mol N}$$

Step 4: Write the empirical formula

C_5H_7N

MOLECULAR FORMULA CALCULATION

But is this the formula for nicotine?

The only way of determining the actual formula of a substance is to know the molecular formula mass.

If the molecular formula mass for nicotine is 162 amu, is the above formula the formula for nicotine?

The mass of C_5H_7N is 81.0 amu.

To determine the molecular formula for nicotine:

$\frac{162}{81.0} = 2$ Therefore, the formula is $C_{10}H_{14}N_2$

CHEMISTRY UNIT 6 PRACTICE PROBLEMS

ATOMIC MASS / MOLECULAR MASS / FORMULA MASS / MOLAR MASS

1. Determine the atomic mass, formula mass, or molecular mass for each of the following (indicate if it is either a(n) atomic mass, formula mass, or molecular mass):
   a. $BaSO_4$ ______________________________

   b. N_2H_4 ______________________________

   c. aluminum oxalate ______________________________

   d. acetic acid ______________________________

   e. arsenic ______________________________

MOLAR MASS

1. Determine the molar mass for each of the following:

   a. ZnI_2 ______________________________

   b. NO_2 ______________________________

   c. Iron(II) nitrate ______________________________

MOLE CONVERSIONS

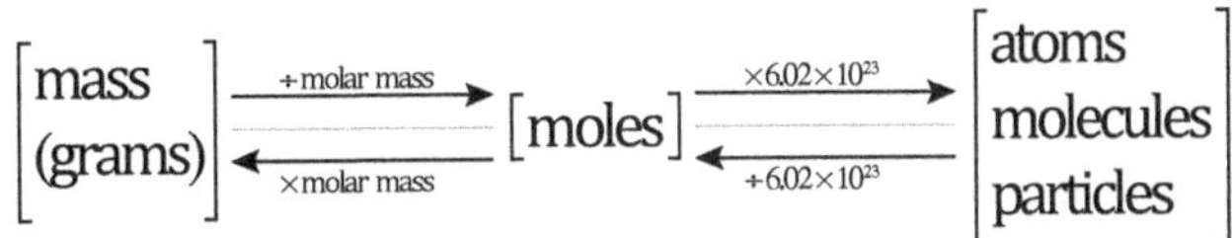

1. Convert mass (grams) to moles.
   Example: Determine the amount of moles in 40.0 grams Mg.
   Set up:

Try:

4.50 grams Fe	0.632 grams C_6H_6
1.6×10^{12} grams Rb	52.670 grams barium chloride
32.12 grams ammonium sulfide	15.6 grams $(NH_4)_2SO_4$

2. Convert moles to mass (grams).
   Example: Determine the mass of 6.43 moles C_2H_4
   Set up:

Try:

15.67 moles Cr	0.00032 moles CO_2
4.56×10^{12} moles S	35,067 moles beryllium nitrate
32,000 moles beryllium nitride	2.00 moles Ne

3. Convert moles to atoms, molecules or particles.
   Example: Determine the number of particles in 0.677 moles $NaHCO_3$
   Set up:

Try:

5.6×10^3 moles Ba	7.64×10^{-1} moles dihydrogen monoxide
0.4428 moles $CaCO_3$	4,560 moles cobalt (II) dichromate
8.21 moles XeF_4	2.00 moles NO

4. Convert atoms, molecules, or particles to moles.

Example: Determine the mole quantity in of 5.6×10^{24} atoms Ag.

Set up:

Try:

4.98×10^{23} particles $MoBr_2$	4.88×10^{21} atoms Ra
2.3333×10^{24} atoms Si	5.1×10^{24} molecules H_2O_2
5.882×10^{26} molecules P_5S_7	7×10^{23} particles MgO

5. Convert mass (grams) to atoms, molecules, or particles.

Examples: How many molecules of alcohol are there in a "standard" can of beer if there are 21.3 g of C_2H_6O?

Set up:

Try:

35.899 grams Au	0.333 grams NH_4OH
54.5 grams Na_2Se	65.6 grams At_2
5.650 grams $SrSO_4$	4.500 grams calcium bromide

6. Convert atoms, molecules, or particles to grams.

Example: What is the mass of 5.888×10^{25} molecules C_3H_8?

Set up:

Try:

4.56×10^{24} atoms Ag	5.6000×10^{20} molecules $C_6H_{12}O_6$
1.230×10^{25} particles BeO	7.2×10^{14} particles $Sn(CrO_4)_2$
3.20×10^{32} atoms Hg	1.54×10^{22} atoms Po

PERCENT COMPOSITION

1. Determine the percentage composition of the elements in each of the following compounds:
   a. barium arsenate ______________________

   b. CH_3COOH ______________________

   c. Ammonium carbonate ______________________

EMPIRICAL AND MOLECULAR FORMULAS

1. Determine the empirical formula for each of the following:
   a. CH_3COOH ______________________

   b. $C_6H_{12}O_6$ ______________________

   c. $N_2O_{10}F_5$ ______________________

2. Determine the empirical formula for each of the following:

30.44% N, 69.55% O, molecular mass = 92 amu
9.93% C, 58.64% Cl, 31.43% F, molecular mass = 121 amu
40.00% C, 6.71% H, 53.29% O, molecular mass = 60 amu

MOLE CALCULATIONS

Tin

87.54 grams Sn convert to moles

Table sugar

0.143 mole $C_{12}H_{22}O_{11}$ convert to grams

Mothballs

1.35×10^{23} molecules $C_{10}H_8$ (naphthalene) convert to moles

Ice-melting salts

0.156 mole $CaCl_2$ convert to particles

Baking soda

23.0 grams $NaHCO_3$ convert to particles

Water

2.5×10^{24} molecules of water convert to grams

REVIEW CALCULATIONS

1. Calculate the amount of molecules in 35.5 grams citric acid (acid found in citrus fruits) $C_6H_8O_7$.

2. What is the mass of 4.59×10^{29} particles of MgO?

3. Is sodium nitrate a molecule or a particle? Explain.

4. Calculate the formula mass of ammonium phosphate.

5. Calculate the molecular mass of C_2H_4OH.

6. Which of the following compound is the term "molecular mass" better suited than the term "formula mass"?
    a. F_2
    b. CaO
    c. K_3N
    d. NaOH
    e. RbOH

7. Calculate the number of moles in 456.8 grams of $Al(OH)_3$.

8. Calculate the molar mass of N_2S

9. What is the mass of 0.0490 mole of titanium(IV) oxide?

10. How many atoms are in 18.8mol Rb?

11. Which of the following statements is incorrect?
    a. Molecular mass is expressed in amu.
    b. Formula mass is expressed in amu.
    c. Atomic mass is expressed in grams/atom.
    d. Molar mass is expressed in grams/mole.

12. Calculate the mass of 6.78×10^{31} particles $Mg_3(PO_4)_2$.

13. How many molecules are in 20.35 mol NH_3?

14. How many atoms are in 44.3 moles of barium?

15. Calculate the number of atoms in 566 grams of carbon.

16. How many grams are in 54.87 mol of CaH_2?

17. How many moles are in 6.27×10^{24} particles of $NaNO_3$?

18. Calculate the mass of 7.99×10^{25} particles Fe_2O_3.

19. Find the percent composition of ammonium nitrate.

20. NutraSweet is 57.14% C, 6.16% H, 9.52% N, and 27.18% O. Calculate the empirical formula of NutraSweet and find the molecular formula. (The molar mass of NutraSweet is 294.30 g/mol.)

ATOMIC, FORMULA, AND MOLECULAR FORMULAS & MASSES

1. Determine the formula or molecular masses for the following compounds:
   a. $Mg(OH)_2$
   b. calcium hydroxide
   c. acetic acid

2. The controversial artificial sweetener saccharin has the molecular formula $C_3H_5O_3NS$. What is its molecular mass?
   a. 123.12 amu
   b. 119.88 amu
   c. 135.14 amu
   d. 103.15 amu
   e. 77.78 amu

3. Cisplatin, an anticancer drug, has the molecular formula $Pt(NH_3)_2Cl_2$. What is the molecular mass of cisplatin?
   a. 323.3 amu
   b. 300.1 amu
   c. 332.6 amu
   d. 321.2 amu
   e. 340.4 amu

4. Nitroglycerin is $C_3H_5N_3O_9$. What is the molecular mass of nitroglycerin?
   a. 240.22 amu
   b. 286.44 amu
   c. 227.10 amu
   d. 270.42 amu
   e. 256.20 amu

5. The formula mass of calcium hydroxide, $Ca(OH)_2$ is:
   a. 128 amu
   b. 74 amu
   c. 97 amu
   d. 57 amu

6. The formula mass of magnesium hydroxide, $Mg(OH)_2$ is:
   a. 42.33 amu
   b. 58.33 amu
   c. 41.32 amu
   d. 5 amu
7. The total number of OXYGEN atoms in the formula of aluminum dichromate, $Al_2(Cr_2O_7)_3$ is:
   a. 10
   b. 7
   c. 29
   d. 21
8. How many atoms are in 18 molecules of glucose, $C_6H_{12}O_6$?
   a. 24
   b. 432
   c. 3240
   d. 1.08×10^{25}
   e. 2.60×10^{26}
9. Which of the following samples contains the smallest number of molecules?
   a. 1 g phosphorus, P_4
   b. 1 g chlorine, Cl_2
   c. 1 g nitrogen, N_2
   d. 1 g arsenic, As_4
   e. 1 g sulfur, S_8

MOLAR MASS

1. The molar mass of sodium chloride, NaCl is:
   a. 58.44 g/mol
   b. 69.71 g/mol
   c. 2 g/mol
   d. 6.022×10^{23} g/mol

THE MOLE CALCULATIONS

1. What is the mass in grams of 10. moles of ammonia, NH_3?
   a. 170 grams
   b. 27.0 grams
   c. 1.70 grams
   d. 0.59 grams
2. $BaSO_4$ is given as a thick slurry before X-rays are taken of the intestinal tract. How many grams are in a 0.568 mole sample of $BaSO_4$?
   a. 62.4
   b. 103
   c. 56.8
   d. 77.8
   e. 133

3. Which of the following correctly describes the mole?
   a. One mole is 6.022×10^{23} atoms of any element.
   b. One mole is the number of atoms in exactly 12.0 g of ^{12}C.
   c. One mole of any chemical compound is one mole of its chemical formula unit.
   d. All of the above are correct.

4. About how many atoms of helium would be found in 2 grams of helium?
   a. 6×10^{23}
   b. 4
   c. 3×10^{23}
   d. 2

5. What is the mass of 4 moles of hydrogen molecules (H_2)?
   a. 4 grams
   b. 3 grams
   c. 8 grams
   d. 1 gram

6. What is the mass in grams of 3.00 moles of water molecules, H_2O?
   a. 54.0 grams
   b. 21.0 grams
   c. 6.01 grams
   d. 0.166 grams

7. How many moles of water molecules, H_2O, are present in a 42.0 gram sample of water?
   a. 2.33 moles
   b. 0.429 moles
   c. 23.98 moles
   d. 757 moles

8. What is the mass of 1.004×10^{23} molecules of barium iodide?
   a. 44.05 g
   b. 44.12 g
   c. 65.20 g
   d. 0.167 g

9. How many moles are in 32.0 grams of CH_4?
   a. 32.0 moles
   b. 16.0 moles
   c. 1.00 mole
   d. 2.00 moles

10. How many moles of methane molecules, CH_4, are in 80 grams of methane?
   a. 1284 moles
   b. 6×10^{80} moles
   c. 0.2 moles
   d. 5 moles

11. The amount of substance having 6.022×10^{23} of any kind of chemical unit is called a(n):
   a. mole
   b. mass number
   c. atomic weight
   d. formula

12. Determine the mass of one mole of CO_2.

13. Determine the number of atoms in one molecule of Fe_2O_3.

14. Determine the number of atoms in one mole of $C_6H_{12}O_6$.

15. Determine the mass of 5.240 moles of gold.

16. Determine the number of moles of nitrogen gas in 85.4 g of nitrogen gas.

17. Determine the mass, in grams, of one atom of silver.

18. Determine the mass, in grams, of one molecule of carbon dioxide.

19. How many molecules are there in 10.0 g of sodium chloride?
    a. 58.4×10^{23}
    b. 5.84×10^{24}
    c. 6.02×10^{24}
    d. 1.03×10^{23}

20. In 0.250 moles of ethylene glycol (antifreeze), $HOCH_2CH_2OH$, there are:
    a. 1.51×10^{23} atoms
    b. 1.51×10^{24} molecules
    c. 1.51×10^{24} atoms
    d. 6.02×10^{24} atoms
    e. 3.01×10^{24} molecules

21. Which of the following does not describe 56.0 g of butene, C_4H_8?
    a. One mole of butene
    b. The amount of butene that contains 8.0 g of hydrogen
    c. The amount of butene that contains $8 \times 6.02 \times 10^{23}$ hydrogen atoms
    d. The amount of butene that contains 48.0 g of carbon
    e. $56.0 \times 6.02 \times 10^{23}$ molecules of butene

22. Sodium cyclamate, $C_6H_{11}NHSO_3Na$, is used as an artificial sweetener in South Africa. If $C_6H_{11}NHSO_3Na$ has a molar mass of 201.2 g/mol, how many moles of sodium cyclamate are contained in a 25.6 g sample?
    a. 0.127 mol
    b. 0.193 mol
    c. 0.245 mol
    d. 7.90 mol
    e. 5180 mol

23. How many moles of nitrogen gas (N_2 molecules) are present in 48.0 grams of nitrogen?
    a. 0.58 mol
    b. 0.86 mol
    c. 1.71 mol
    d. 2.00 mol
    e. 3.42 moles

24. Which one of the following has the lightest mass?
    a. An HF molecule
    b. 20.0 g of HF
    c. 10.0 mol of H_2
    d. 1 mol of F_2
    e. 1 mol of H_2O

25. One mole is _____.
    a. the amount of molecules in any substance
    b. the amount of particles in any substance
    c. the amount of atoms in any substance
    d. the amount of ions in any substance
    e. just a number

PERCENT COMPOSITION

1. Determine the percent composition of ammonium sulfate, $(NH_4)_2SO_4$.

2. Determine the percent composition of urea, N_2H_4CO.

3. Each of the compounds listed in questions 1 & 2 contains nitrogen. They are used as fertilizers. For each of the compounds, which one has the highest percentage of nitrogen?

4. Calculate the percentage composition of $Ca(ClO_3)_2$?
    a. 19.4% Ca, 34.3% Cl, 46.4% O
    b. 32.4% Ca, 28.7% Cl, 38.8% O
    c. 49.0% Ca, 21.8% Cl, 29.3% O
    d. 32.4% Ca, 67.6% ClO_3
    e. 19.4% Ca, 51.0% ClO_3

EMPIRICAL & MOLECULAR FORMULAS

1. Write the empirical formula for each of the following compounds.
   a. $C_{12}H_{22}O_{11}$, sugar

   b. $C_2H_6O_2$, ethylene glycol (antifreeze.

2. From the following empirical formulas and the formula masses for each compound, determine their molecular formulas.
   a. CH_3; formula mass = 30.0 amu

   b. CH_2; formula mass = 84.0 amu

   c. $C_3H_4O_3$ (Vitamin C); formula mass = 176 amu

3. Determine the empirical formula for a compound that contains 18.6 grams of phosphorus and 14.0 grams of nitrogen.

4. Determine the empirical formula for a compound that contains 35.6% of phosphorus and 64.4% of sulfur.

5. A compound with a molecular mass of 98.0 g/mole was determined to be 24.49% carbon, 4.08% hydrogen, and 72.43% chlorine.
   a. Determine the empirical formula of the compound.

   b. Determine the molecular formula of the compound.

6. What is the empirical formula for N_8O_4?
   a. NO_2
   b. NO
   c. N_2O
   d. ON_4
   e. N_2O_2

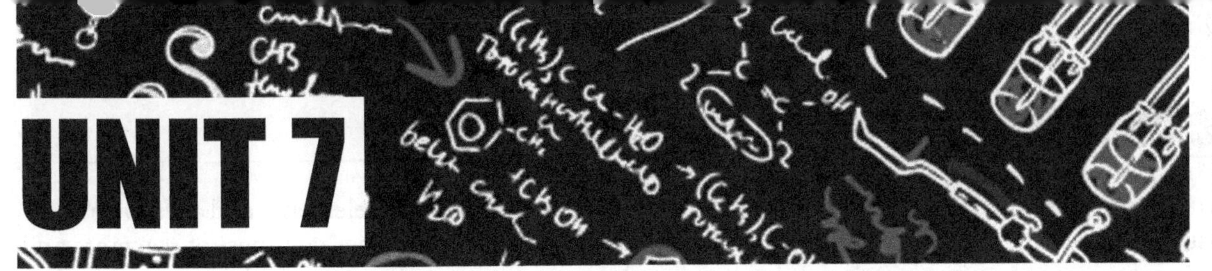

REACTIONS—STOICHIOMETRY IS A BIG WORD (REACTIONS AND EQUATIONS)

Stoichiometry—The Meaning of the Word

Evolution of a Chemical Equation

The Chemical Reaction

Balancing Chemical Equations

Classifying Chemical Reactions

Conversion Factors from a Chemical Equation

Stoichiometry—History

Review: Chemical Reactions/Equations

- Stoichiometry Problems
  - Mole A → Mole C
  - Mole B → Gram C
  - Gram C → Mole A
  - Gram C → Gram A
  - Gram C → Gram A (Another Approach)
- Percentage Yield
- Limiting Reactant Problems

OBJECTIVES FOR UNIT 7

- To be able to write and balance an unbalanced chemical equation.
- To be able to write balanced equations for combination, decomposition, oxidation—reduction, and double replacement reactions.
- To use solubility rules to help with some general rules for water solubility's of common ionic compounds.
- To understand a chemical equation and the relationships between reactants and products.
- To convert mole and mass amounts in a chemical equation (stoichiometry).
- To perform percentage yield calculations.
- To perform limiting reactant problems.

STOICHIOMETRY—THE MEANING OF THE WORD

The word stoichiometry derives from two Greek words: *stoicheion* (meaning element) and *metron* (meaning measure). Stoichiometry deals with calculations about the masses (sometimes volumes) of reactants and products involved in a chemical reaction. It is a very mathematical part of chemistry, so be prepared for lots of calculator use.

EVOLUTION OF A CHEMICAL EQUATION

As a result of Dalton's atomic theory, many compounds were being investigated and given chemical formulas. A Swedish scientist named Jöns Jakob Berzelius is credited with presenting the law of multiple proportions which states, "In a series of compounds made up of the same element, a simple ratio exists between the weights of one and the fixed weight of the other element." Berzelius continued his investigating in the 1790s, eventually devised a simpler system for all the chemical symbols and names being used for compounds.

© Nicku/Shutterstock.com

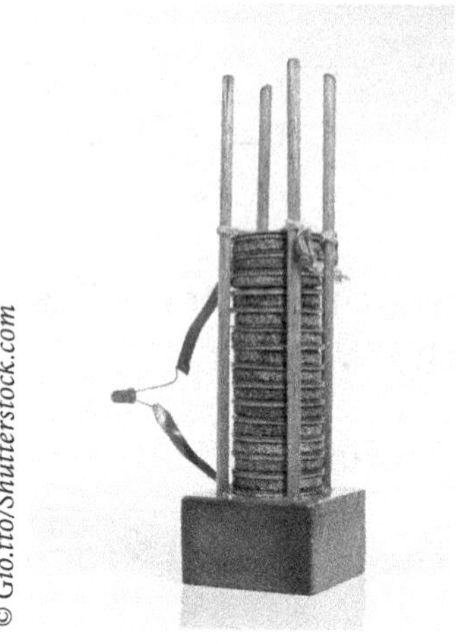
© Gio.tto/Shutterstock.com

Berzelius used the voltaic pile (battery) to divide compounds. He proposed that metals always went to the negative pole and nonmetals to the positive pole of the "electrical machine." Ben Franklin also studied this "electrical machine" in which he propose the idea of + and – electricity. This work leads us to a more systematic representation of chemical changes (and physical changes in some cases).

THE CHEMICAL REACTION

There are three basic steps involved in writing a chemical equation:

1. Decide which substances are the reactants and which are the products.

Magnesium burns (reacting it with oxygen) producing magnesium oxide

© Wooden Owl/Shutterstock.com

2. Determine the correct formula for each substance and substitute it for the name.

$Mg(s) + O_2(g) \rightarrow MgO(s)$

3. Balance the chemical equation.

$2\ Mg(s) + O_2(g) \rightarrow 2\ MgO(s)$

Terms and symbols: *(aq)* aqueous *(s)* solid *(g)* gas *(l)* liquid

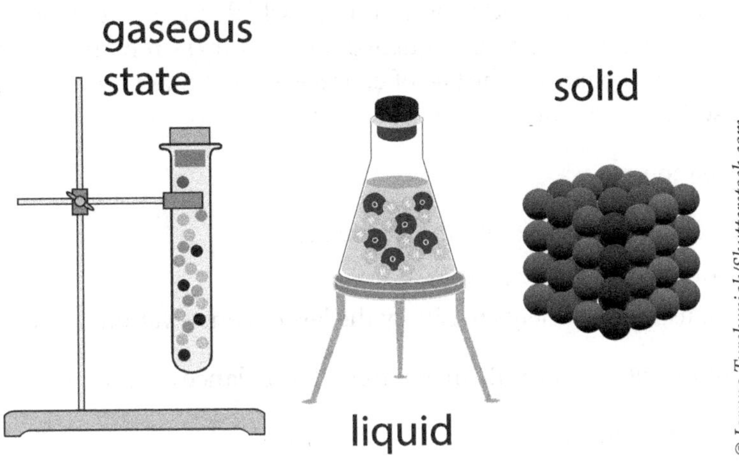

© Iaryna Turchyniak/Shutterstock.com

Other examples:

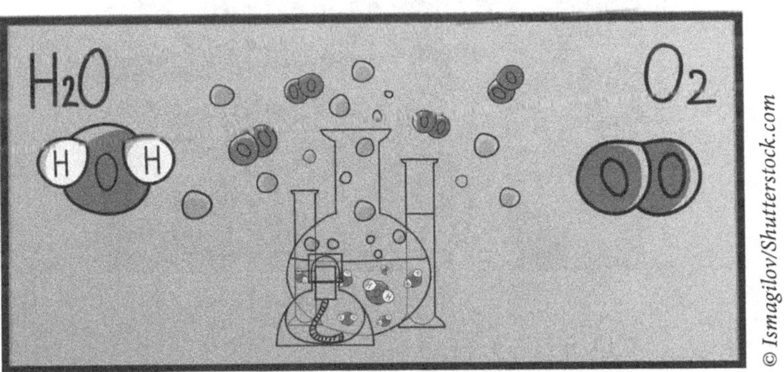

© Ismagilov/Shutterstock.com

$2H_2(g) + O_2(g) \rightarrow 2H_2O(l)$

reactants *products*

- Reactants: 2 mol $H_2(g)$ and 1 mol $O_2(g)$
- Products: 2 mol $H_2O(l)$

Zinc combines with hydrochloric acid producing zinc chloride and a gas

Equation:

$$Zn(s) + 2\ HCl(g) \rightarrow ZnCl_2(s) + H_2(g)$$

© Andrei Marincas/Shutterstock.com

BALANCING CHEMICAL EQUATIONS

In Unit 3 we learned about the law of conservation of mass, which states: Atoms can neither be created nor destroyed in a chemical reaction, and as a consequence the total mass remains unchanged. When material undergoes physical and chemical changes, the law of conservation of mass is represented in a chemical equation. This is achieved by writing a balanced chemical equation.

To balance a chemical equation:

1. Determine exactly what are the reactants and products.
2. Assemble the parts of the chemical equation.
3. Balance the equation using coefficients to satisfy the law of conservation of mass.

Note: never change the subscript in a formula in an attempt to balance an equation.

The reaction of propane burning in a furnace:

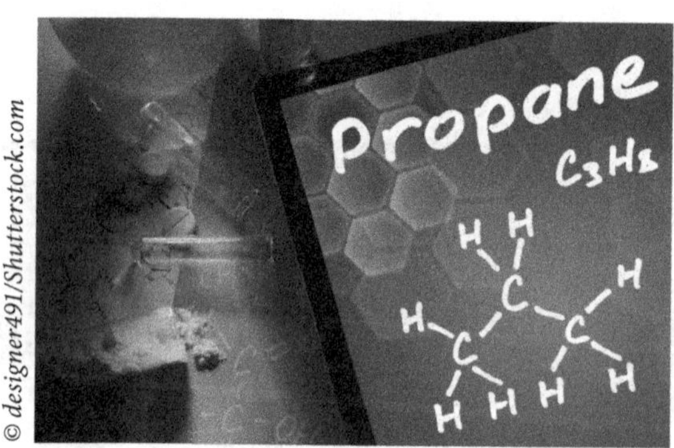

© designer491/Shutterstock.com

$$\_\_\_\ C_3H_8 + \_\_\_\ O_2 \rightarrow \_\_\_\ CO_2 + \_\_\_\ H_2O$$

3	C	1
8	H	2
2	O	3

Balanced:

$C_3H_8 + 5\ O_2 \rightarrow 3\ CO_2 + 4\ H_2O$

3	C	3
8	H	8
10	O	10

The chemical used to "chlorinate" a swimming pool.

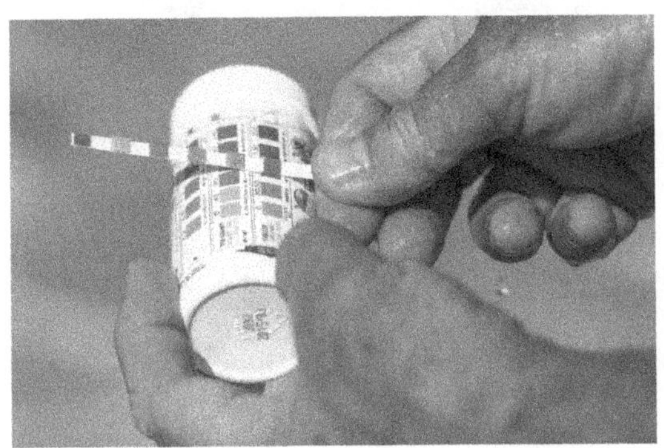

© pixinoo/Shutterstock.com

$\_\_\_\_ Cl_2 + \_\_\_\_ NaOH \rightarrow \_\_\_\_ NaOCl + \_\_\_\_ NaCl + \_\_\_\_ H_2O$

2	Cl	2
1	Na	2
1	O	2
1	H	2

Balanced:

$Cl_2 + 2\ NaOH \rightarrow NaOCl + NaCl + H_2O$

2	Cl	2
2	Na	2
2	O	2
2	H	2

CLASSIFYING CHEMICAL REACTIONS

Chemical changes are represented with chemical reactions. As we investigate these changes, we find that similarities in many substances and the way that they react. As a result, we can classify chemical reactions into different categories.

1. **Combination (synthesis) reactions** involve two or more substances combining to form one substances.

© microvector/Shutterstock.com

In the form: A + Z → AZ

$Zn(s) + Cl_2(l) \rightarrow ZnCl_2(s)$

2. **Decomposition reactions** involve a substance breaking up into smaller substances.

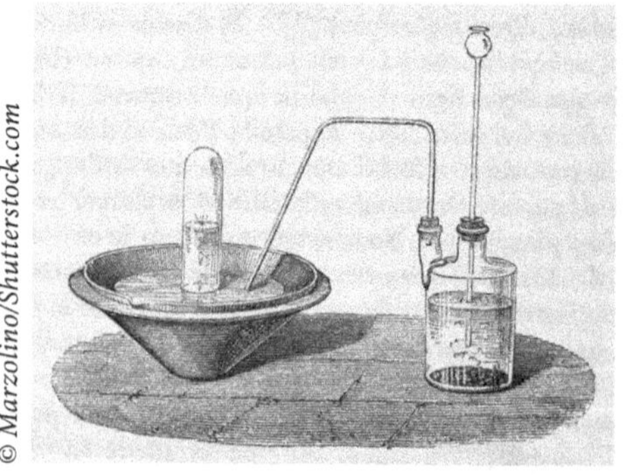

© Marzolino/Shutterstock.com

In the form: AZ → A + Z

$2HgO(s) \rightarrow 2\ Hg(l) + O_2(g)$

3. **Combustion reactions** involve a reaction that involves oxygen as a reactant.

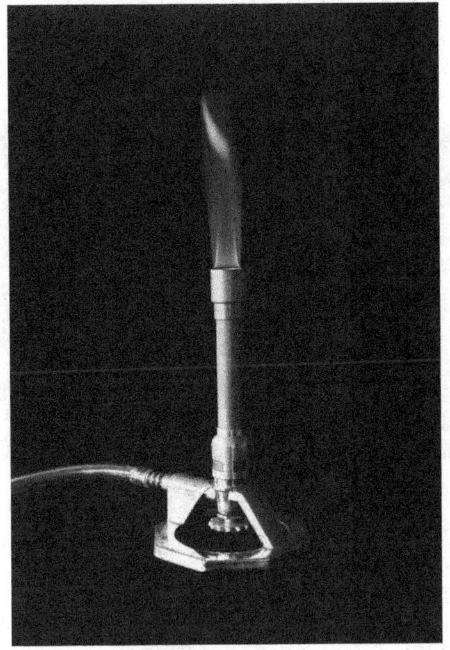

© ggw1962/Shutterstock.com

$C_3H_8(g) + 5O_2(g) \rightarrow 3CO_2(g) + 4H_2O(g)$

4. **Single replacement reactions** (oxidation—reduction reactions) involve one element displacing another element in a compound.

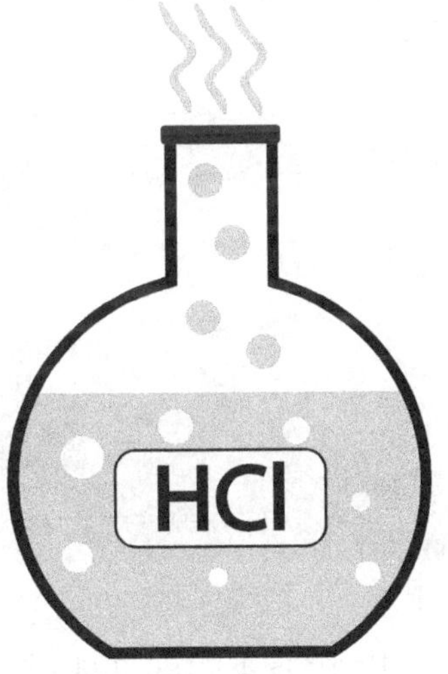

© konstantinks/Shutterstock.com

In the form: A + BZ → AZ + B

$Zn(s) + 2HCl(aq) \rightarrow H_2(g) + ZnCl_2(aq)$

Compounds in our environment often undergo oxidation-reduction reactions (redox reactions), which involve the redistribution of electrons.

© Eldred Lim/Shutterstock.com

Examples of redox reactions:

All types of batteries such alkaline, NiCad, car batteries; rusting and corrosion; metabolism

1. **Double replacement reactions** "positive" and "negative" interchanging

In the form: $AX + BZ \rightarrow AZ + BX$

$$AgNO_3(aq) + Na_2CrO_4(aq) \rightarrow AgCrO_4(s) + 2\ NaNO_3(aq)$$

a. **Precipitate reactions** (still double replacement reactions)

Precipitate is the solid product that may possibly form. Often it can be recovered by filtering.

Soluble: An ionic substance that breaks apart into separate ions when in solution. Soluble substances often "appear" to have disappeared.

Insoluble: An ionic substance that does not break apart in solution and stays as a solid substance. This is the product that forms the precipitate.

Example: Magnesium in seawater. Because there is a large quantity of magnesium in seawater, using the understanding of soluble and insoluble, we can remove the magnesium by forming the precipitate $Mg(OH)_2$.

Molecular equation:

$$2NaOH(aq) + MgCl_2(aq) \rightarrow Mg(OH)_2(s) + 2\ NaCl(aq)$$

Note: Many precipitate reactions are double replacement reactions.

After working with different solutions in a variety of double replacement reactions, we are able to determine the products that are formed. From the products formed, we can determine if soluble or insoluble substances are formed. We can refer to solubilty rules for further examples.

$Pb(NO_3)_2(aq) + 2\ NaI(aq) \rightarrow PbI_2(s) + 2\ NaNO_3(aq)$

© Lindsey Moore/Shutterstock.com

$CuSO_4(aq) + BaCl_2(aq) \rightarrow BaSO_4(s) + CuCl_2$

b. **Neutralization** reactions (still double replacement)

In the form: Acid + Base → Salt + Water

$HCl(aq) + NaOH(aq) \rightarrow NaCl(aq) + H_2O(l)$

In aqueous solution, HCl and NaOH exist as ions in solution. The salt, NaCl, that forms is soluble in water; therefore, it also exists as ions in solution.

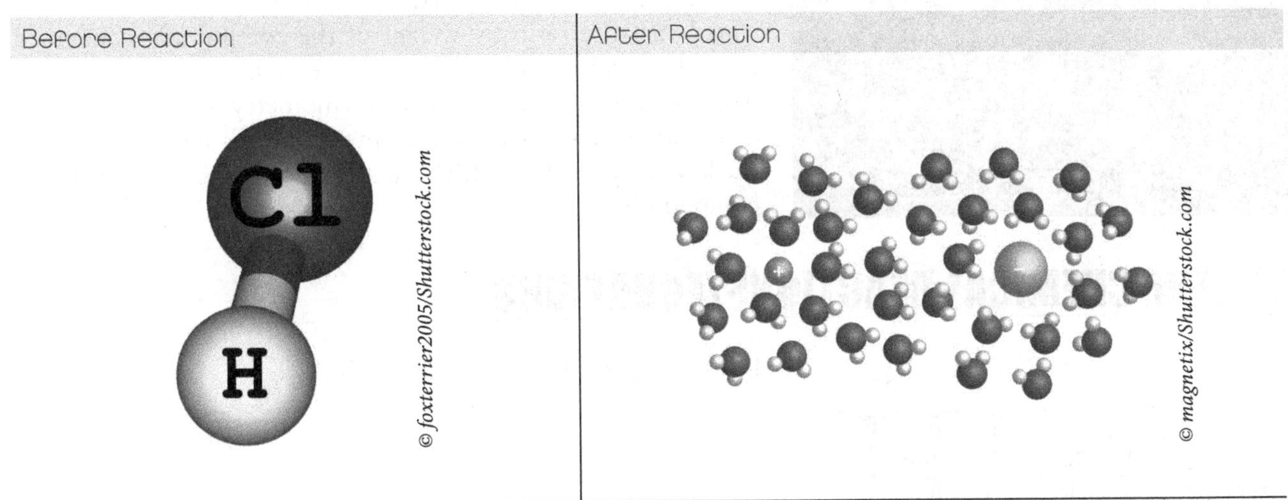

© foxterrier2005/Shutterstock.com
© magnetix/Shutterstock.com

Example: Dolomitic lime cuts wastewater treatment costs by reducing the amount of alkali necessary for neutralization. It offers a 23% reduction in product use versus hydrated calcium lime. When neutralizing sulfuric acid-based wastewater, the sludge volume generated can be cut in half.

Neutralization Comparison

For every ton of sulfuric acid neutralized with high-calcium lime, 1.4 dry tons of calcium sulfate are formed. The calcium sulfate and entrapped water require disposal.

$$2Ca(OH)_2 + 2H_2SO_4 \rightarrow \underset{(1.4\text{ tons})}{2CaSO_4} + 4H_2O$$

When dolomitic lime is substituted for high-calcium lime in this reaction, the amount of calcium sulfate generated is reduced by 50%.

$$Ca(OH)_2MgO + 2H_2SO_4 \rightarrow \underset{(0.7\text{ tons})}{CaSO_4} + MgSO_4 + 3H_2O$$

The magnesium sulfate formed in this reaction is soluble, and will not require disposal. This represents a dramatic saving in landfill costs.

CONVERSION FACTORS FROM A CHEMICAL EQUATION

STOICHIOMETRY—HISTORY

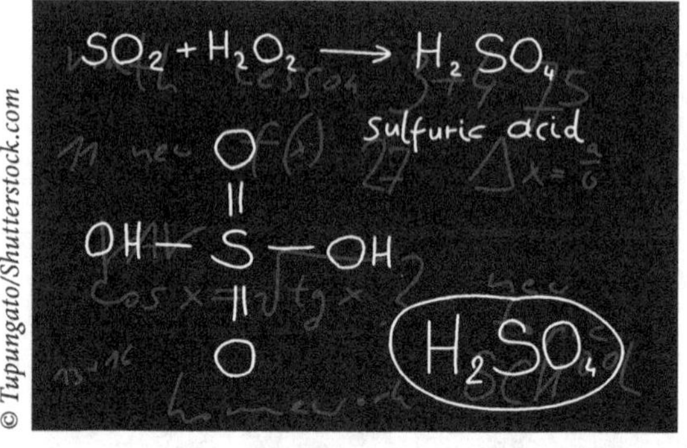

Jeremias Benjamin Richter, a German chemist, became a mining official in 1794. He performed some of the earliest determinations of the quantities by weight in which acids saturate bases. He proposed the law of definite proportions which stated that the ratio by weight of the compounds consumed in a chemical reaction is always the same. In his published works he introduced the term stoichiometry, which he defined as the art of chemical measurements, which has to deal with the laws according to which substances unite to form chemical compounds.

REVIEW: CHEMICAL REACTIONS/EQUATIONS

Reactants: $Zn + I_2 \rightarrow$ Product: $Zn\ I_2$

Chemical Equations

stoichiometric coefficients

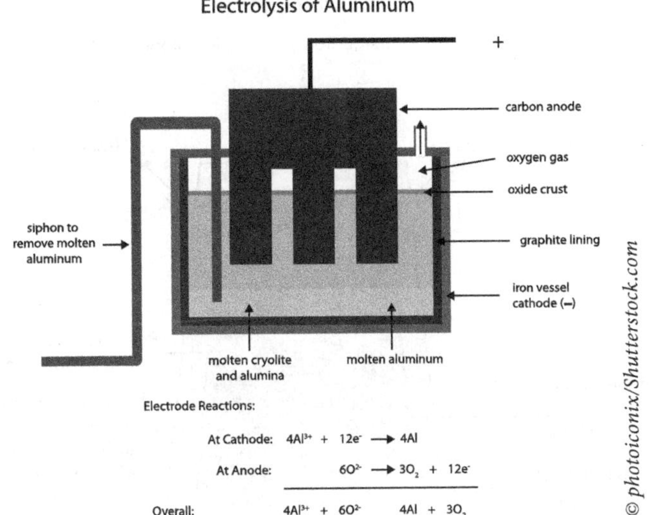

$2\ Al_2O_3 \rightarrow 4\ Al + 3\ O_2$

This equation means:

2 particles aluminium oxide → 4 atoms aluminium + 3 molecules oxygen

2 moles aluminium oxide → 4 moles aluminium + 3 moles oxygen

Why is this important?

As fascinating as it was to find out that based upon the atomic masses on the periodic table we can determine the amount of atoms, molecules, and particles our material has, chemical stoichiometry is just as amazing!

Before we let elements and compounds meet, we can predict how they are going to react, and even further, how much stuff they are going to produce. We could also be on the other end of things and find a waste product and determine exactly how much of a particular compound or element was needed to produce the waste. How wonderful! This is **stoichiometry**!

Reaction stoichiometry involves the mass and mole relationships among reactants and products in a chemical reaction. A balanced chemical equation is used to determine these relationships.

Note: The stoichiometric coefficients in a balanced equation give the relative numbers of atoms, or molecules, or the number of moles of each substance reacting.

STOCHIOMETRIC PROBLEMS

There are four basic stoichiometry problems to learn:

$$A + B \rightarrow C$$

MOLE A → MOLE C

A reaction with iron oxide:

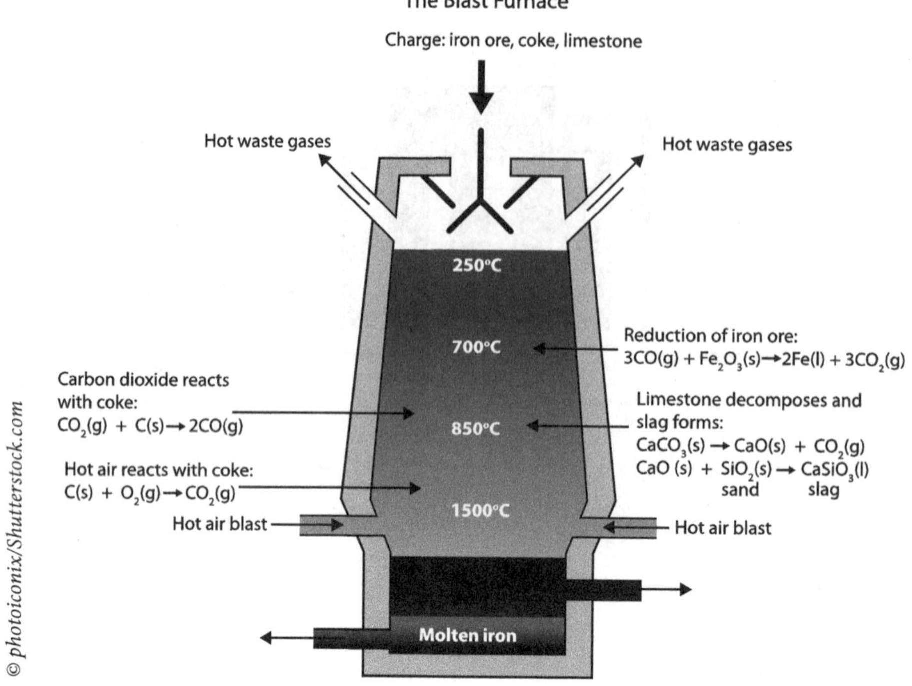

© photoiconix/Shutterstock.com

$$3\ CO(g) + Fe_2O_3(s) \rightarrow 2\ Fe(l) + 3\ CO_2(g)$$

How many moles of carbon monoxide are needed to react to produce 15.65 moles of iron?

Step 1: convert mol Fe to mol carbon monoxide (bridge step)

The chemical equation indicates:

$$3 \text{ mol } CO(g) + 1 \text{ mol } Fe_2O_3(s) \rightarrow 2 \text{ mol } Fe(s) + 3 \text{ mol } CO_2(g)$$

$$15.65 \text{mol Fe} \times \frac{3 \text{mol CO}}{2 \text{mol Fe}} = 23.48 \text{mol CO}$$

MOLE B → GRAM C

A reaction with aluminum and chlorine:

© Artgraphixel/Shutterstock.com

$$2\ Al + 3\ Cl_2 \rightarrow Al_2Cl_6$$

How many grams of Al_2Cl_6 are produced when 0.144 moles of chlorine are reacting?

Step 1: convert moles Cl_2 to moles Al_2Cl_6 (bridge step)

The chemical equation indicates:

$2 \text{ mol Al} + 3 \text{ mol } Cl_2 \rightarrow 1 \text{ mol } Al_2Cl_6$

$$0.144 \text{mol } Cl_2 \times \frac{1 \text{mol } Al_2Cl_6}{3 \text{mol } Cl_2} = 0.0480 \text{mol } Al_2Cl_6$$

Step 2: convert moles Al_2Cl_6 to grams Al_2Cl_6 (mol→g conversion, multiply by molar mass)

$$0.0480 \text{mol } Al_2Cl_6 \times \frac{267.0 \, Al_2Cl_6}{1 \text{mol } Al_2Cl_6} = 12.8 \text{g } Al_2Cl_6$$

GRAM C → MOLE A

A reaction of hydrogen and oxygen:

© Marynchenko Oleksandr/ Shutterstock.com

$2 H_2 + O_2 \rightarrow 2 H_2O$

If 45,450 grams of oxygen are put into the fuel cells to react with enough hydrogen, how many moles of water are produced?

Step 1: convert grams O_2 to moles O_2 (g→mol conversion, divide by molar mass)

$$45{,}450 \text{g } O_2 \times \frac{1 \text{mol } O_2}{32.0 \text{g } O_2} = 1420.3 \text{mol } O_2$$

Step 2: convert moles O_2 to moles H_2O (bridge step)

$2 \text{ mol } H_2 + 1 \text{ mol } O_2 \rightarrow 2 \text{ mol } H_2O$

$$1420.3 \text{mol } O_2 \times \frac{2 \text{mol } H_2O}{1 \text{mol } O_2} = 2840.6 \text{mol } H_2O$$

GRAM C → GRAM A

A reaction of decomposing potassium chlorate:

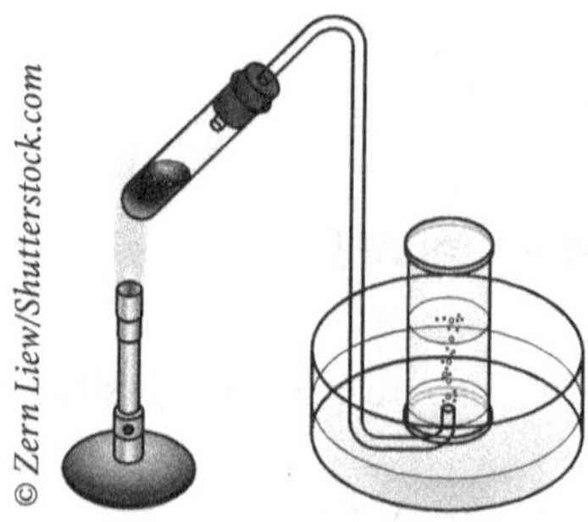
© Zern Liew/Shutterstock.com

$2\ KClO_3 \rightarrow 2\ KCl + 3\ O_2$

What mass of potassium chloride is produced with 53.4 grams oxygen when potassium chlorate decomposes?

Step 1: convert grams O_2 to moles O_2 (g→mol conversion, divide by molar mass)

$$53.4\,g\ O_2 \times \frac{1\,mol\ O_2}{32.0\,g\ O_2} = 1.67\,mol\ O_2$$

Step 2: convert moles O_2 to moles KCl (bridge step)

The chemical equation indicates:

$2\ mol\ KClO_3 \rightarrow 2\ mol\ KCl + 3\ mol\ O_2$

$$1.67\,mol\ O_2 \times \frac{2\,mol\ KCl}{3\,mol\ O_2} = 1.11\,mol\ KCl$$

Step 3: convert moles KCl to grams KCl (mol→g conversion, multiply by molar mass)

$$1.11\,mol\ KCl \times \frac{74.6\,g\ KCl}{1\,mol\ KCl} = 83.0\,g\ KCl$$

GRAM C → GRAM A (another approach)

A reaction of decomposing potassium chlorate:

$2\ KClO_3 \rightarrow 2\ KCl + 3\ O_2$

What mass of potassium chloride is produced with 53.4 grams oxygen when potassium chlorate decomposes?

Step 1: convert balanced equation to a gram relationship (mol→g conversion)

The chemical equation indicates:

$2\ mol\ KClO_3 \rightarrow 2\ mol\ KCl + 3\ mol\ O_2$
[122.6 g/mol] [74.6 g/mol][32.0 g/mol]

The converted chemical equation:

$245.2\ g\ KClO_3 \rightarrow 149.2\ g\ KCl\ +\ 96.0\ g\ O_2$

Step 2: convert moles O_2 to moles KCl (bridge step with gram relationship)

$$53.4\,g\,O_2 \times \frac{149.2\,g\,KCl}{96.0\,O_2} = 83.0\,g\,KCl$$

PERCENTAGE YIELD

Experimentally, quantities of products are determined that may not always match those quantities of products determined through stoichiometric calculations. The percentage of product obtained experimentally can be determined by percentage yield.

$$\text{Percentage yield} = \frac{\text{Actual yield}}{\text{Theoretical yield}}$$

Percent yield tells the efficiency of the reaction.

Actual yield is the amount of product really obtained (by experimentation).

Theoretical yield is the amount of product that should be obtained (by calculation).

A little history, to illustrate a reaction of incandescent light bulbs.

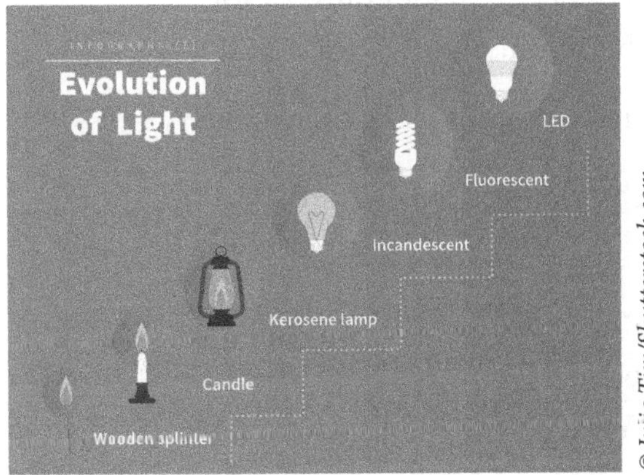

© Julia Tim/Shutterstock.com

It's a little-known fact that if African American Lewis Latimer hadn't been on Thomas Edison's team, the light bulb might have been too impractical for most people to use. Edison's light bulb used bamboo as the filament and the light bulb burned out after only 30 hours. Latimer, along with Joseph V. Nichols, came up with both the idea to use carbon filaments and the process for manufacturing the carbon filaments. Carbon lasted much longer and made the light bulb practical. At present, tungsten metal, W, is used to make incandescent bulb filaments.

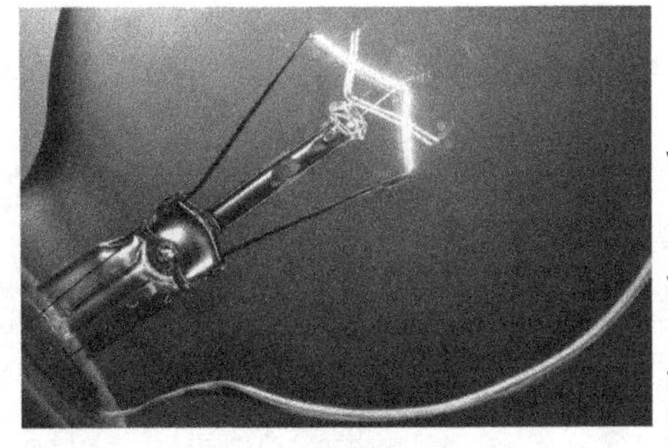

© Chones/Shutterstock.com

The metal is separated out of yellow tungsten (VI) oxide, WO_3 by reacting with hydrogen according to the following reaction: $WO_3 + 3\,H_2 \rightarrow W + 3\,H_2O$

If 2.76 grams of W is experimentally obtained from the reaction between 15.8 grams of WO_3 and excess of H_2, calculate the percentage yield obtained from the reaction.

First, determine the theoretical yield of W using stoichiometry

$$231.8 \text{ g } WO_3 + 6.0 \text{ g } H_2 \rightarrow 183.8 \text{ g W} + 54.0 \text{ g } H_2O$$

$$15.8\text{g } WO_3 \times \frac{183.8\text{g W}}{231.8\text{g } WO_3} = 12.5\text{g W}$$

Second, determine the percentage yield of W

$$\text{Percent Yield} = \frac{2.73\text{g W}}{12.5\text{g W}} \times 100 = 22.0\%$$

Thus, only 22.0% of the tungsten was recovered.

© vvoe/Shutterstock.com

Another example: Copper can be recovered from a mineral in our environment: Cu_2S (called chalcocite). When I performed this experiment, I started with 15.0 grams of Cu_2S and recovered 9.95 grams of copper. What was the percentage yield of copper from my experiment?

Write the chemical equation:

$$Cu_2S \rightarrow 2\text{ Cu} + \text{S}$$
$$\text{mol } Cu_2S \rightarrow 2\text{ mol Cu} + 1\text{ mol S}$$
$$159.1\text{ g } Cu_2S \rightarrow 127\text{ g Cu} + 32.1\text{ g S}$$

First, determine the theoretical yield of copper using stoichiometry.

$$15.0\text{g } Cu_2S \times \frac{127\text{g Cu}}{159.1\text{g } Cu_2S} = 12.0\text{g Cu}$$

Second, determine the percentage yield of W

$$\text{Percent Yield} = \frac{\text{Actual Yield}}{\text{Theoretical Yield}} \times 100$$

$$\text{Percent Yield} = \frac{9.95\text{g}}{12.0\text{g}} \times 100 = 82.9\%\text{ Cu}$$

Thus, only 82.9% copper was recovered.

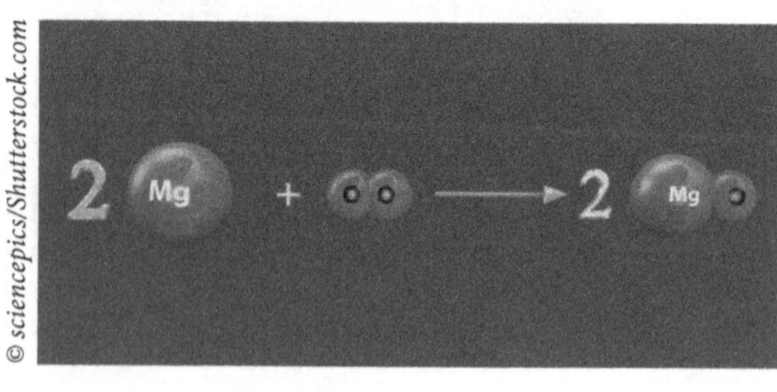

© sciencepics/Shutterstock.com

LIMITING REACTANT PROBLEMS

Limiting reactant is the reactant that is completely used up in a chemical reaction. It limits the amount of other reactants that can combine and the amount of products formed in a reaction.

Excess reactant is the reactant in excess in a chemical reaction. It is not completely used up in a reaction.

Why is this important?

In our age of environmental awareness, we try to minimize our waste production. The same is true in chemical processes. We learned with stoichiometry that we can determine the exact amount of product that will be produced when using a known amount of reactant. In many chemical reactions more than one reactant is used. How can we minimize using too much of our reactants to produce a certain amount of product?

Remember the following things about limiting reactants:

- The reaction will stop when the reactants are used up.
- If one reactant is used up before the other, the reaction will stop.
- The first reactant used up is the limiting reactant; use it for the calculation.
- The other reactant is the excess reactant.

Example:

Mix 5.40 g of Al with 8.10 g of Cl_2. What mass of Al_2Cl_6 can form?

$2\ Al + 3\ Cl_2 \rightarrow Al_2Cl_6$
$2\ mol\ Al + 3\ mol\ Cl_2 \rightarrow 1\ mol\ Al_2Cl_6$
$2\ Al + 3\ Cl_2 \rightarrow Al_2Cl_6$

Convert g Al to g Al_2Cl_6 and convert g Cl_2 to g Al_2Cl_6.

$$5.40 g\ Al \times \frac{1 mol\ Al}{27.0 g\ Al} = 0.200 mol\ Al$$

$$8.10 g\ Cl_2 \times \frac{1 mol\ Cl_2}{71.0 g\ Cl_2} = 0.114 mol\ Cl_2$$

Step 2: (bridge step) convert mol Al to mol Al_2Cl_6 and convert mol Cl_2 mol Al_2Cl_6.

The reactant that produced the smaller mol Al_2Cl_6 is the L.R.

$$0.200 mol\ Al \times \frac{1 mol\ Al_2Cl_6}{2 mol\ A1} = 0.100 mol\ Al_2Cl_6$$

$$0.114 mol\ Cl_2 \times \frac{1 mol\ Al_2Cl_6}{3 mol\ Cl_2} = 0.0380 mol\ Al_2Cl_6$$

Cl_2 is the limiting reactant and Al is the excess reactant.

Step 3: Calculate mass of Al_2Cl_6 expected, using the limiting reactant.

$$0.0380 mol\ Al_2Cl_6 \times \frac{267.0 g\ Al_2Cl_6}{1 mol\ Al_2Cl_6} = 10.1 g\ Al_2Cl_6$$

How much of excess reactant will remain when reaction is complete?

Step 1: Determine the amount of excess reactant (Al) is actually needed in the reaction.

$$0.114 mol\ Cl_2 \times \frac{2 mol\ Al}{3 mol\ Cl_2} = 0.0760 mol\ Al$$

$$0.0760 mol\ Al \times \frac{27.0 g\ Al}{1 mol\ Al} = 2.05 g\ Al$$

Step 2: Subtract this amount from the amount of Al the reaction started with.

5.40 g Al – 2.05 g Al = 3.35 g Al is in excess when the reaction is complete.

Additional example: The fuel in the main cells of the Space Shuttle are liquid oxygen and liquid hydrogen.

The explosive reaction results in the production of H_2O. Obviously, those working on the space shuttle had to have an understanding of limiting reactant problems.

The reaction: $2\ H_2 + O_2 \rightarrow 2\ H_2O$

If 45,450 grams of oxygen and 45,620 grams of hydrogen where put into the fuel cells of the space shuttle.

Which reactant is the limiting reactant?

$$45{,}450\text{g } O_2 \times \frac{1\text{mol } O_2}{32.0\text{g } O_2} \times \frac{2\text{mol } H_2O}{1\text{mol } O_2} = 2{,}841\text{mol } H_2O$$

$$45{,}620\text{g } H_2 \times \frac{1\text{mol } H_2}{2.0\text{g } H_2} \times \frac{2\text{mol } H_2O}{2\text{mol } H_2} = 22{,}810\text{mol } H_2O$$

Oxygen is the limiting reactant.

How much H_2O will be produced?

$$2{,}841\text{mol } H_2O \times \frac{18.0\text{g } H_2O}{1\text{mol } H_2O} = 51{,}140\text{g } H_2O$$

How much excess reactant did not need to be put into the space shuttle?

$$45{,}450\text{g } O_2 \times \frac{1\text{mol } O_2}{32.0\text{g } O_2} \times \frac{2\text{mol } H_2}{1\text{mol } O_2} \times \frac{2.0\text{g } H_2}{1\text{mol } H_2} = 5{,}681\text{g } H_2$$

45,620g H_2 – 5,681 g H_2 = 39,940 g H_2 in excess.

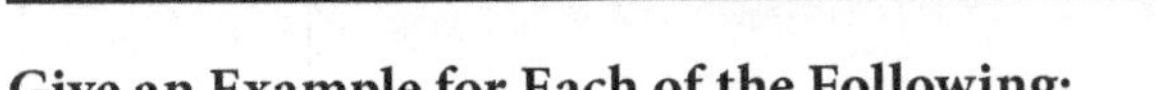

CHEMISTRY UNIT 7 PRACTICE PROBLEMS

Give an Example for Each of the Following:

Combination (CA) Decomposition (D) Combustion (CU)

Single Replacement (SR) Double Replacement (DR)

Chemical Reactions	Classify
1. Magnesium burning Write the chemical reaction. Identify the reactants and products. Identify the phase labels. Balance the equation	
2. copper + silver nitrate → silver + copper (II) nitrate Write the chemical reaction. Identify the reactants and products. Identify the phase labels. Balance the equation	
3. Explain at the particulate (in molecule ratio) and the macroscopic (in mole ratio), what the coefficients represent in the combination of hydrogen and oxygen to make water.	
4. ____ $Mg(s)$ + ____ $ZnCl_2(aq)$ → ____ $MgCl_2(aq)$ + ____ $Zn(s)$	
5. ____ $AgNO_3(aq)$ + ____ CaCl2(aq) → ____ $AgCl(s)$ + ____ $Ca(NO_3)_2(aq)$	
6. ____ $C_2H_6(g)$ + ____ $O_2(g)$ → ____ $CO_2(g)$ + ____ $H_2O(g)$	
7. ____ $Na_2O(s)$ + ____ $H_2O(l)$ → ____ $NaOH(aq)$	
8. ____ $KClO_3(s)$ → ____ $KCl(s)$ + ____ $O_2(g)$	
9. ____ $Al(s)$ + ____ $HCl(aq)$ → ____ $AlCl_3(aq)$ + ____ $H_2(g)$	
10. ____ $C_3H_6O(g)$ + ____ $O_2(g)$ → ____ $CO_2(g)$ + ____ $H_2O(g)$	

Chemical Reactions	Classify
11. ____ $Fe(s)$ + ____ $O_2(g)$ → ____ $Fe_2O_3(s)$	
12. ____ $Cl_2(aq)$ + ____ $KBr(aq)$ → ____ $Br_2(aq)$ + ____ $KCl(aq)$	
13. ____ $Ca(NO_3)_2(aq)$ + ____ $K_3PO_4(aq)$ → ____ $Ca_3(PO_4)_2(s)$ + ____ $KNO_3(aq)$	
14. ____ $Ca(HCO_3)_2(aq)$ → ____ $CaCO_3(s)$ + ____ $H_2O(l)$ + ____ $CO_2(g)$	
15. ____ $NaOH(aq)$ + ____ $H_3PO_4(aq)$ → ____ $Na_3PO_4(s)$ + ____ $H_2O(l)$	
16. ____ $BaCl_2(aq)$ + ____ $Na_2SO_4(aq)$ → ____ $BaSO_4(s)$ + ____ $NaCl(aq)$	
17. ____ $Fe(s)$ + ____ $CuCl_2(aq)$ → ____ $Cu(s)$ + ____ $FeCl_2(aq)$	
18. ____ $C_6H_6(l)$ + ____ $O_2(g)$ → ____ $CO_2(g)$ + ____ $H_2O(l)$	
19. ____ $BaCl_2(s)$ + ____ $H_2O(l)$ → ____ $BaCl_2 \bullet H_2O(s)$	
20. ____ $Pb(NO_3)_2(aq)$ + ____ $K_2CrO_4(aq)$ → ____ $PbCrO_4(s)$ + ____ $KNO_3(aq)$	
21. ____ $Mg(s)$ + ____ $H_2SO_4(aq)$ → ____ $MgSO_4(aq)$ + ____ $H_2(g)$	
22. ____ $CaO(s)$ + ____ $H_2O(l)$ → ____ $Ca(OH)_2(s)$	
23. ____ $C_2H_4O(l)$ + ____ $O_2(g)$ → ____ $CO_2(g)$ + ____ $H_2O(l)$	
24. ____ $HNO_3(aq)$ + ____ $Ba(OH)_2(aq)$ → ____ $Ba(NO_3)_2(aq)$ + ____ $H_2O(l)$	

STOICHIOMETRY

THE DECOMPOSITION OF HYDROGEN PEROXIDE (H_2O_2)

1. What do you think the bubbles or gas are?

2. What do you think causes the bubbling when hydrogen peroxide is placed on an open wound?

	Reactants	→	Products	
equation		→		+
mole ratio		→		+
molar mass				
gram ratio		→		+

INVESTIGATION OF THE SUGAR MOLECULE

1. The molecule below is a sucrose or sugar molecule. Use lines to connect each water molecule to prove that there are 11 water molecules in sucrose.

Sucrose (saccharose)

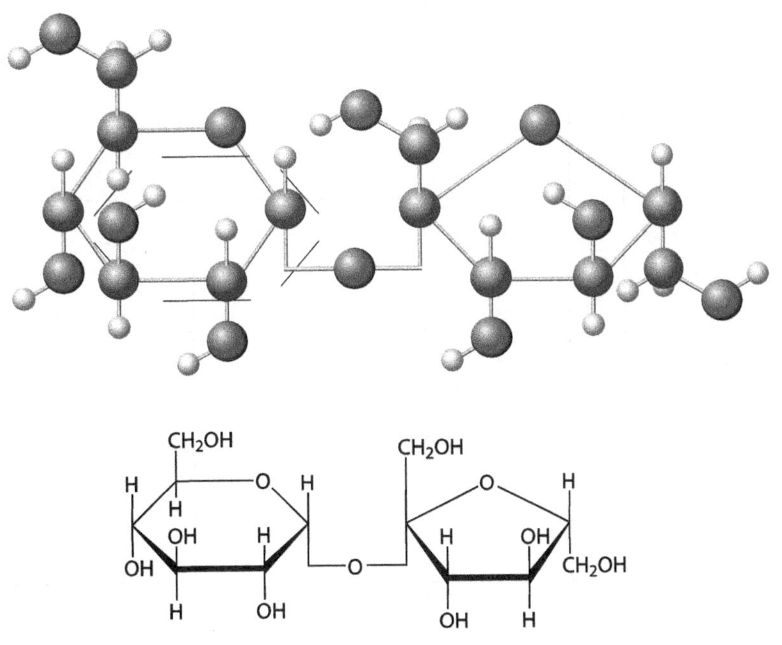

2. The molecule below is a simple carbohydrate called glucose or blood sugar. What is the formula for glucose?

3. Place circles around hydrogen and oxygen atoms, then use lines to connect 2 hydrogen and 1 oxygen atoms to prove there are 6 water molecules.

4. Prove that sucrose is made up of two glucose-like molecules minus one H_2O molecule.

α-D-Glucose (cyclic)

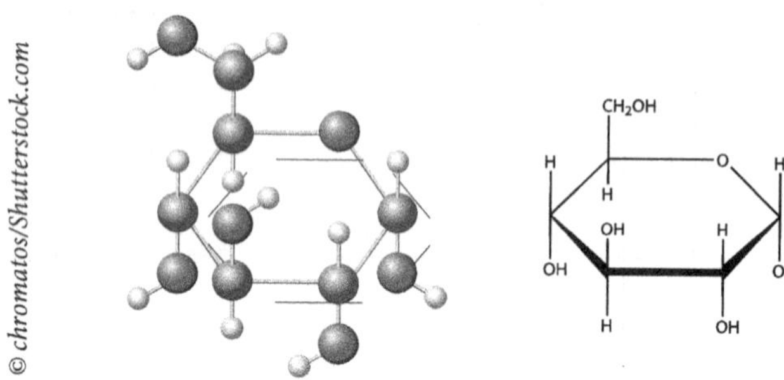

Data Table: Decomposition of sucrose (demonstrated)

	Reactants	→	Products	
equation		→		+
mole ratio		→		+
molar mass				
gram ratio		→		+

STOICHIOMETRY

A + B → C

MOLE A → MOLE C

A reaction with iron oxide:

$Fe_2O_3(s) + 3H_2(g) \rightarrow 2Fe(s) + 3H_2O(l)$

How many moles of hydrogen are needed to react to produce 15.65 moles of iron?

Now you try:

1. Combustion of Propane (C_3H_8).

   $C_3H_8(g) + 5O_2(g) \rightarrow 3CO_2(g) + 4H_2O(g)$

   Calculate the number of moles of oxygen reacting if 4.56 moles of CO_2 is produced.

2. How many moles of oxygen are consumed when 32.4 moles of carbon dioxide is formed during the combustion of C_7H_{16}?

   $C_7H_{16} + 11O_2 \rightarrow 7CO_2 + 8H_2O$

3. How many moles of $C_6H_{12}O_6$ are formed when 0.250 mole of CO_2 is consumed in the reaction

   $6CO_2 + 6H_2O \rightarrow C_6H_{12}O_6 + 6O_2$

MOLE B → GRAM C

A reaction with aluminum and chlorine:

$2Al + 3Cl_2 \rightarrow Al_2Cl_6$

How many grams of Al_2Cl_6 are produced when 0.144 moles of chlorine are reacting?

Now you try:

1. Combustion of Propane (C_3H_8).

   $C_3H_8(g) + 5O_2(g) \rightarrow 3CO_2(g) + 4H_2O(g)$

   What is the mass of propane reacting if 765 moles of water is produced?

2. In the combustion of natural gas according to the equation

   $CH_4 + 2O_2 \rightarrow CO_2 + 2H_2O$

   how many grams of water are formed during the combustion of 0.264 mole of CH_4?

3. In the equation: $I_2 + 7F_2 \rightarrow 2IF_7$

   3.28 moles of fluorine will yield ______ grams of IF_7.

GRAM C → MOLE A

A reaction of hydrogen and oxygen:

$2H_2 + O_2 \rightarrow 2H_2O$

If 45,450 grams of oxygen where put into the fuel cells to react with enough hydrogen, how many moles of water are produced.

Now you try:

1. Combustion of Propane (C_3H_8).

   $C_3H_8(g) + 5O_2(g) \rightarrow 3CO_2(g) + 4H_2O(g)$

   How many moles of carbon dioxide are produced from 56.8 grams of oxygen reacting?

2. In the complete combustion of $C_3H_8O_3$, how many moles of carbon dioxide are produced when 23.0 g of $C_3H_8O_3$ burns?

   $2C_3H_8O_3 + 7O_2 \rightarrow 6CO_2 + 8H_2O$

3. How many moles of bromine will react with 6.50 grams of C_2H_2 in the reaction

   $C_2H_2 + 2Br_2 \rightarrow C_2H_2Br_4$

GRAM C → GRAM A

A reaction of decomposing potassium chlorate:

$2KClO_3 \rightarrow 2KCl + 3O_2$

What mass of potassium chloride is produced with 53.4 grams oxygen when potassium chlorate decomposes?

Now you try:

1. Combustion of Propane (C_3H_8).

   $C_3H_8(g) + 5O_2(g) \rightarrow 3CO_2(g) + 4H_2O(g)$

   What is the mass of propane reacting to produce 3.43 grams of water?

2. According to the following equation:

   $4NH_3 + 5O_2 \rightarrow 4NO + 6H_2O$

   How many grams of NO will form when 40. grams of oxygen are used?

3. In the equation:

   $N_2 + 2O_2 \rightarrow 2NO_2$

   64 grams of oxygen will yield _____ grams of nitrogen dioxide.

STOCHIOMETRY PROBLEMS

1. How many moles of chlorine gas (Cl_2) would react with 5.0 moles of sodium (Na) according to the following chemical equation? (Balance equation.)

   _____ Na + _____ Cl_2 → _____ NaCl

2. If you start with 10.0 grams of aluminum hydroxide, how many grams of water will be produced?

   $Al(OH)_3 + 3HBr \rightarrow AlBr_3 + 3H_2O$

3. If you start with 4.500 moles of ethylene (C_2H_4), how many grams of carbon dioxide will be produced?

   $C_2H_4 + 3O_2 \rightarrow 2CO_2 + 2H_2O$

4. If you start with 5.5 grams of NaF, how many moles of magnesium fluoride will be produced?

   $Mg + 2NaF \rightarrow MgF_2 + 2Na$

5. If you start with 20.0 grams of hydrochloric acid, how many grams of sulfuric acid will be produced?

   $2HCl + Na_2SO_4 \rightarrow 2NaCl + H_2SO_4$

6. $C_3H_8 + 5O_2 \rightarrow 3CO_2 + 4H_2O$

   a. If I start with 5.0 grams of C_3H_8, what is my theoretical yield of water?

   b. Calculate the percentage yield if 6.24 grams of water are actually obtained by experiment (with starting with 5.0 grams C_3H_8).

7. My theoretical yield of beryllium chloride was 10.7 grams. If my actual yield was 4.5 grams, what was my percentage yield?

   $Be + 2HCl \rightarrow BeCl_2 + H_2$

BALANCE CHEMICAL EQUATIONS

1. Balance the following chemical equations.
   a. $CH_4 + O_2 \rightarrow CO_2 + H_2O$
   b. $La_2(CO_3)_3 + H_2SO_4 \rightarrow La_2(SO_4)_3 + H_2O + CO_2$
   c. $Al(NO_3)_3 + Na_2CO_3 \rightarrow Al_2(CO_3)_3 + NaNO_3$

2. According to the following unbalanced reactions:
   a. Balance the equations
   b. Determine the amount of reactants and products (in moles)
   c. Explain the phase labels

   i. $KClO_4(s) \rightarrow KCl(s) + O_2(g)$

   ii. $BaCl_2(aq) + (NH_4)2CO_3(aq) \rightarrow BaCO_3(s) + NH_4Cl(aq)$

   iii. aluminum metal reacts with phosphoric acid to produce hydrogen gas and solid aluminum phosphate.

3. What is the coefficient for H_2 when the equation $Ba + H_3AsO_4 \rightarrow H_2 + Ba_3(AsO_4)_2$ Is properly balanced?
   a. 1
   b. 3
   c. 5
   d. 2
   e. 4

4. Calcium combines with bromine to make calcium bromide. Write the balanced chemical equation for the reaction. What is the coefficient for bromine?
   a. 1
   b. 2
   c. 3
   d. 4
   e. 5

5. According to the following reaction:
   $2Mg(s) + O_2(g) \rightarrow 2MgO(s)$
   What is the phase of the product?
   a. solid
   b. gas
   c. solid and gas
   d. liquid
   e. liquid and gas

6. Barium peroxide, BaO_2, breaks down into barium oxide and oxygen. Write the balanced chemical equation for this reaction. What is the coefficient for barium oxide?
   a. 3
   b. 1
   c. 5
   d. 6
   e. 2

7. Lithium combines with oxygen to form lithium oxide. Write the balanced chemical equation for this reaction. What is the coefficient for lithium?
   a. 4
   b. 3
   c. 5
   d. 1
   e. none of the above

8. Interpret the following sentence:
   "Sodium bicarbonate reacts with acetic acid to produce sodium acetate, carbon dioxide and water."
   a. $Na_2CO_3 + H_2C_2H_3O_2 \rightarrow H_2 + CO_3 + Na_2C_2H_3O_2$
   b. $NaCO_3 + HC_2H_3O_2 \rightarrow NaHC_2H_3O_2 + CO_2 + H_2O$
   c. $NaHCO_3 + H_2O + CO_2 \rightarrow NaC_2H_3O_2 + 2O_2$
   d. $NaHCO_3 + HC_2H_3O_2 \rightarrow NaC_2H_3O_2 + H_2O + CO_2$

CLASSIFY CHEMICAL EQUATIONS

1. Classify the following chemical reactions as:

combination	decomposition	single replacement
double replacement	neutralization	combustion

   a. $Fe(s) + 2HCl \rightarrow H_2(g) + FeCl_2$
   b. $4NH_3 + 5O_2(g) \rightarrow 4NO + 6H_2O$
   c. $C_2H_2 + HCl \rightarrow C_2H_3Cl$
   d. $NaBr(aq) + AgNO_3(aq) \rightarrow AgBr(s) + NaNO_3(aq)$
   e. $C_6H_{12}O_6 \rightarrow 2C_2H_5OH + 2CO_2$
   f. $NaOH + CH_3COOH \rightarrow NaCH_3COO + H_2O$

2. The decomposition by heating of solid potassium chlorate yields solid potassium chloride and oxygen gas as products. Write a balanced equation for this reaction.
   a. $KClO_4 \rightarrow KCl(s) + 2O_2(g)$
   b. $2KClO_3(s) \rightarrow 2\ KCl(s) + 3O_2(g)$
   c. $KClO_3(s) \rightarrow KCl(s) + 3O(g)$
   d. $2KClO_3(s) \rightarrow 2KClO_2(s) + O(g)$
   e. $KClO_2(s) \rightarrow KCl(s) + O_2(g)$

3. $2Al + 3Sn(NO_3)_2 \rightarrow 2Al(NO_3)_3 + 3Sn$ This equation is an example of which type of reaction?
   a. single replacement
   b. double replacement
   c. combination
   d. decomposition

4. $6K_2O + P_4O_{10} \rightarrow 4K_3PO_4$ This equation is an example of which type of reaction?
   a. single replacement
   b. double replacement
   c. combination
   d. decomposition

5. In class a double displacement reaction was done as a demonstration. A solution of potassium iodide and a solution of lead (II) nitrate were combined. A yellow precipitate formed as a product. What was the precipitate?
   a. KI
   b. $Pb(NO_3)_2$
   c. KNO_3
   d. PbI_2
   e. KPb

6. $2H_{2(g)} + CO_{(g)} \rightarrow CH_3OH_{(l)}$ This equation is an example of which type of reaction?
   a. single replacement
   b. double replacement
   c. combination
   d. decomposition

7. Based upon the type of reaction, determine the products. Write the balanced chemical equation.
   a. Combination: Calcium and bromine combine.

   b. Decomposition: Water breaks apart by electrolysis.

   c. Combustion: Natural gas (CH_4) is burned in furnaces.

   d. Single Replacement: Chlorine gas is bubbled through an aqueous solution of potassium iodide.

   e. Double Replacement: Calcium nitrate and potassium fluoride combine to form a precipitate.

   f. Double Replacement: Neutralization: Sodium hydroxide is added to phosphoric acid.

STOCHIOMETRY MOLE TO MOLE CONVERSIONS

1. Balance the following equation: _____ CCl_4 + _____ SbF_3 → _____ CCl_2F_2 + _____ $SbCl_3$

2. According to the following chemical reaction:
   _____ $Al(s)$ + _____ $H_2SO_4(aq)$ → _____ $Al_2(SO_4)_3(aq)$ + _____ $H_2(g)$
   a. Balance the equation.
   b. Indicate the amount of moles reacting and moles being produced of each reactant and product.
3. According to the following chemical reaction:
   $3CaCO_3(s) + 2H_3PO_4(aq) \rightarrow Ca_3(PO_4)_2(aq) + 3H_2O(l) + 3CO_2(g)$
   a. Calculate the moles of calcium phosphate being produced from 45.6 moles of phosphoric acid reacting.
   b. Calculate the moles of calcium carbonate reacting if 0.344 moles of water are produced.
4. How many moles of carbon dioxide are in the following chemical reaction if 2 moles of C_4H_{10} are reacting?
   $2C_4H_{10}(g) + 13O_2(g) \rightarrow 8CO_2(g) + 10H_2O(g)$

5. How many moles of water are being produced from 4 moles C_4H_{10} and 26 moles of O_2 reacting?
____ $C_4H_{10}(g)$ + ____ $O_2(g)$ → ____ $CO_2(g)$ + ____ $H_2O(g)$

6. How many moles of oxygen are in the following chemical reaction?
____ $Na_2CO_3(aq)$ + ____ $HNO_3(g)$ → ____ $H_2O(l)$ + ____ $CO_2(g)$ + ____ $NaNO_3(aq)$

7. Calculate the number of moles of Fe_3O_4 produced from 0.75 moles of Fe by the following reaction.
____ Fe + ____ H_2O ____ Fe_3O_4 + ____ H_2

8. Calculate the number of moles of NO produced from 0.25 moles of O_2 by the following reaction.
____ NH_3 + ____ O_2 ____ NO + ____ H_2O

STOCHIOMETRY MOLE TO GRAM CONVERSIONS

1. According to the following chemical reaction:
$Al_2O_3(s) + 6HNO_3(aq) \rightarrow 2Al(NO_3)_3(aq) + 3H_2O(l)$
Calculate the mass of nitric acid if 5.60 moles of water are produced.

2. What mass of CaO could be obtained from the thermal decomposition of 2.00 moles of $CaCO_3$?
____ $CaCO_3$ → ____ CaO + ____ CO_2

3. How many grams of aluminum bromide are formed by the reaction of 1.50 moles of HBr according to the following equation?
____ Al + ____ HBr → ____ $AlBr_3$ + ____ H_2

4. How many grams of H_2 are produced by the reaction of 0.256 mol of H_3PO_4 according to the following equation?
____ Cr + ____ H_3PO_4 → ____ $CrPO_4$ + ____ H_2

STOCHIOMETRY GRAM TO MOLE CONVERSIONS

1. According to the following chemical reaction:
   $Al_2O_3(s) + 6HNO_3(aq) \rightarrow 2Al(NO_3)_3(aq) + 3H_2O(l)$
   Calculate the moles of aluminum oxide if 43.3 grams of aluminum nitrate are produced.

STOCHIOMETRY GRAM TO GRAM CONVERSIONS

1. According to the following chemical reaction:
   $Al_2O_3(s) + 6HNO_3(aq) \rightarrow 2Al(NO_3)_3(aq) + 3H_2O(l)$
   a. Calculate the mass of water produced if 0.430 grams of aluminum nitrate are produced.
   b. Calculate the mass of aluminum oxide reacting if 88.8 grams of water are produced.
2. Freshly exposed aluminum surfaces react with oxygen to form a tough oxide coating that protects the metal from further corrosion. How many grams of O_2 are required to react with 8.09 g of Al?
   ____ Al + ____ $O_2 \rightarrow$ ____ Al_2O_3
3. Barium chloride was used to precipitate silver chloride from a solution of silver nitrate. What mass of barium chloride had to react if 0.635 grams of silver chloride formed?
   ____ $BaCl_2(aq)$ + ____ $AgNO_3(aq) \rightarrow$ ____ $AgCl(s)$ + ____ $Ba(NO_3)_2(aq)$

STOCHIOMETRY COMBINED CONVERSIONS

1. According to the following reaction:
   ____ WO_3 + ____ $H_2 \rightarrow$ ____ W + ____ H_2O
   How many moles of tungsten are produced by the reaction of 0.00761 mole of WO_3 with hydrogen?
2. According to the following reaction:
   ____ $Mg(s)$ + ____ $O_2(g) \rightarrow$ ____ $MgO(s)$
   How many moles of MgO are produced of 4.60 moles of oxygen react?
3. According to the following reaction:
   ____ N_2 + ____ $O_2 \rightarrow$ ____ NO
   How many moles of NO produced by the reaction of 26.8 grams of nitrogen?

4. According to the following reaction:

 $\_\_\_\_ CH_4(g) + \_\_\_\_ O_2(g) \rightarrow \_\_\_\_ CO_2(g) + \_\_\_\_ H_2O(g)$

 How many moles of NO produced by the reaction of 26.8 grams of nitrogen?

5. According to the following reaction:

 $\_\_\_\_ Al(s) + \_\_\_\_ O_2(g) \rightarrow \_\_\_\_ Al_2O_3(s)$

 How many grams of oxygen react with 108 grams aluminum?

6. According to the following reaction:

 $\_\_\_\_ CCl_4 + \_\_\_\_ SbF_3 \rightarrow \_\_\_\_ CCl_2F_2 + \_\_\_\_ SbCl_3$

 If 4.36 grams of Freon-12 (CCl_2F_2) is produced in the reaction, how many grams of $SbCl_3$ are also produced?

STOICHIOMETRY LIMITING REACTANT

1. If 25.0 grams of Al_2O_3 and 75.0 grams of carbon react according to the following chemical reaction:

 $Al_2O_3(s) + 3C(s) \rightarrow 2Al(s) + 3CO(s)$

 a. Which is the limiting reactant and which is the excess reactant?
 b. What is the maximum number of grams of Al that can be produced?
 c. What is the mass of excess reactant remaining after the reaction?

2. According to the following reaction:

 $\_\_\_\_ Sb + \_\_\_\_ Cl_2 \rightarrow \_\_\_\_ SbCl_3$

 If Sb is completely consumed in the reaction, it is the (limiting or excess) reactant?

3. What mass of $PbSO_4$ is produced when 1.94 g $Pb(NO_3)_2$ reacts with 0.83 g $Al_2(SO_4)_3$?

 $\_\_\_\_ Pb(NO_3)_2 + \_\_\_\_ Al_2(SO_4)_3 \rightarrow \_\_\_\_ PbSO_4 + \_\_\_\_ Al(NO_3)_2$

4. Cu_2HgI_4 is prepared according to the equation:

 $2CuI + HgI_2 \rightarrow Cu_2HgI_2$

 When 2.00 grams of each reactant are used, which one is the limiting reactant?

5. If the reaction $N_2 + 3H_2 \rightarrow 2NH_3$ is carried out using 1.40 grams of N_2 and 0.400 grams of H_2, what mass of excess reactant will remain?

PERCENTAGE YIELD CALCULATIONS

1. According to the following chemical reaction:

   _____ Na_2CO_3 + _____ HCl → _____ NaCl + _____ H_2O + _____ CO_2

   5.00 grams of Na_2CO_3 was treated with excess HCl, 1.50 grams of CO_2 was obtained (actual yield).

   a. Calculate the theoretical amount of grams of CO_2 produced.
   b. Calculate the percentage yield of CO_2.

2. Reaction of 1.00 mole CH_4 with excess Cl_2 yields 96.8 g CCl_4 (actual yield). What is the percentage yield of the reaction?

   _____ CH_4 + _____ Cl_2 → _____ CCl_4 + _____ HCl

3. Toluene is oxidized by air under carefully controlled conditions to benzoic acid which is used to prepare the food preservative sodium benzoate. What is the yield of a reaction in which 1.00 g of toluene is converted to 1.21 g of benzoic acid (actual yield)?

   $2C_6H_5CH_3 + 3O_2 \rightarrow 2C_6H_5CO_2H + 2H_2O$

4. The reaction of 6.8 g of H_2S with excess SO_2 according to the following reaction yields 8.2 g of S (actual yield). What is the percentage yield?

   _____ H_2S + _____ SO_2 → _____ S + _____ H_2O

5. Which is the best percentage yield?

   a. 0%
   b. 10%
   c. 50%
   d. 75%
   e. 100%

6. In a general chemistry laboratory experiment, a student produces 2.73 grams of a compound. She calculates the theoretical yield as 3.40 grams. What is the percentage yield?

PERCENTAGE YIELD WITH LIMITING REACTANT CALCULATIONS

1. The mass of Li_2O formed when 2.00 g of lithium reacts with 2.00 g of oxygen is 3.02 g (actual yield). What is the percentage yield?

   _____ Li + _____ O_2 → _____ Li_2O

2. The mass of S_2Cl_2 formed when 6.00 g of sulfur reacts with 6.00 g of chlorine is 9.5 g (actual yield). What is the percentage yield?

   _____ S_8 + _____ Cl_2 → _____ S_2Cl_2

3. The mass of iron produced by the reaction of 7.00 g of Fe_2O_3 and 3.00 g of CO is 3.55 g (actual yield). What is the percentage yield?

   $Fe_2O_3 + 3CO \rightarrow 2Fe + 3CO_2$

4. If 2.75 grams of NaI are produced from a mixture initially containing 5.00 grams I_2 and 1.00 grams of NaOH, what is the percentage yield?

   $3I_2 + 6NaOH \rightarrow 5NaI + NaIO_3 + 3H_2O$

UNIT 8

IT'S A GAS (GAS LAWS)

Gas Molecules

Associations with Gases:

- Pressure (P)
- Volume (V)
- Temperature (T)
- Amount of Substance (n)

Standard Temperature and Pressure (STP)

Gas Laws

- Boyle's Law
- Charles's Law
- Gay-Lussac's Law
- Combined Gas Law
- Avogadro's Law
- Combined Gas Law
- The Ideal Gas Law
    - PV = nRT
    - Molecular Mass Calculations
    - Density Calculations

Gas Stoichiometry

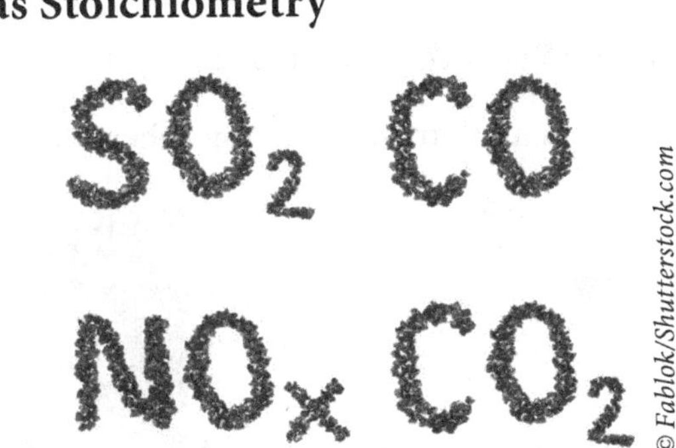

OBJECTIVES FOR UNIT 8

- To learn the characteristics of properties gases.
- To understand the measurements and conversions of pressure, temperature, and volume.
- To learn and calculate with the gas laws the lead to the combined gas law.
- To learn and calculate the ideal gas law.
- To be able to convert between molar mass and volume using the ideal gas law.
- To be able to perform stoichiometric calculations using the gas laws.

GAS MOLECULES

Gases such as hydrogen, oxygen, and nitrogen have properties that change, depending on temperature, pressure, volume, and amount of gas present.

Characteristics of gases:

1. Gases expand indefinitely and uniformly.
2. Gases have no definite shape or volume.
3. Gases can be highly compressed.
4. Gases are of lower densities than liquids and solids.
5. Most non-reacting gases mix uniformly.

ASSOCIATIONS WITH GASES: PRESSURE, VOLUME, TEMPERATURE, AND MOLES

1. **Pressure (P)** is force per unit area.

A barometer is used to measure atmospheric pressure.

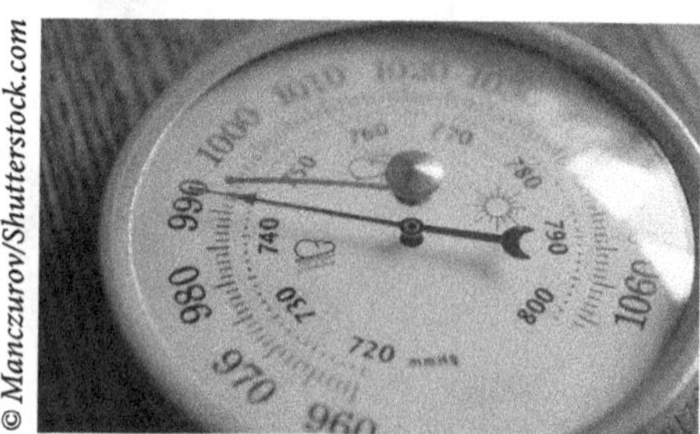

A manometer is used to measure the pressure of a gas (or liquid) in a sealed flask.

Units: $1atm = 760.torr = 760.mmHg = 101.3kPa = 14.7psi$

Example:

Convert the pressure inside a can of pop:

$4560torr$ to _______ atm to _______ $mmHg$ to _______ kPa to _______ psi

$$4560 \times \frac{1.00 atm}{760. torr} = 6.00 atm$$

$$4560 torr \times \frac{1 mmHg}{1 torr} = 4560 mmHg$$

$$4560 torr \times \frac{101.3 kPa}{760. torr} = 608 kPa$$

$$4560 torr \times \frac{14.7 psi}{760. torr} = 88.2 psi$$

Calculate the pressure (in Pascals) of a closed monometer filled with oil. The pressure is 453 mmHg. The density of mercury at 0°C is 13.596 g/mL.

2. **Volume (V)** is the amount of space occupied by a gas.

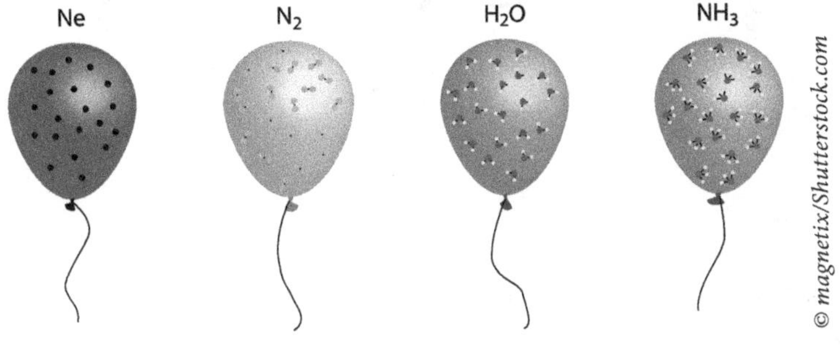

Units: 1L = 1000mL = 1000cm^3 = 1dm^3

3. **Temperature (T)** is a measure of the average kinetic energy of the particles in a sample of gas. (**Note:** K.E. = 1/2 mv^2 m=mass v=velocity)

Units: °F = (1.8 × °C) + 32 K = 273.15 + °C

Example:

Calculate the temperature (in °C and Kelvin) of oxygen at 55°F.

$$^{\circ}C = \frac{^{\circ}F - 32}{1.8} = \frac{55 - 32}{1.8} = 13^{\circ}C$$

K = 273 + 13 = 286 K

4. **Amount of substance (n)** is the amount of gas present in the quantity of moles. At a condition known as STP, 1 mole of any gas will occupy a volume of 22.4 liters. **STP is Standard Temperature (0 degrees Celsius) and Standard Pressure (1 atm).**

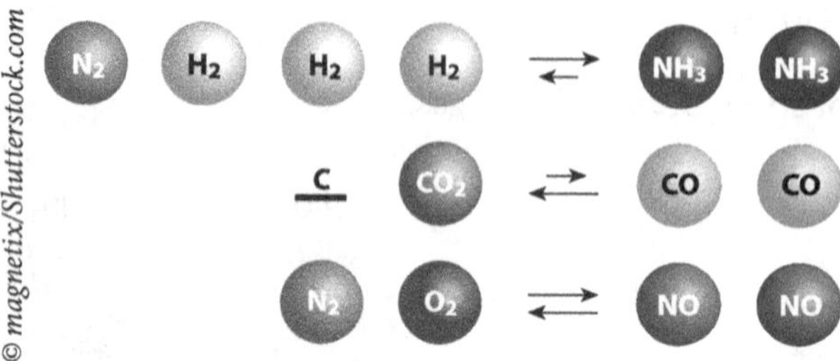

© magnetix/Shutterstock.com

Units: moles

Example: Calculate the volume of 45.6 grams nitrogen dioxide at STP.

$$45.6gNO_2 \times \frac{1molNO_2}{46.0gNO_2} = 0.991molNO_2$$

$$0.991molNO_2 \times \frac{22.4LNO_2}{1molNO_2} = 22.2LNO_2$$

GAS LAWS

BOYLE'S LAW

Robert Boyle, (1627–1691), stated that at constant temperature, the volume of a fixed mass of a given gas is inversely proportional to the pressure it exerts.

At constant temperature: $P_1V_1 = P_2V_2$

© Georgios Kollidas/Shutterstock.com

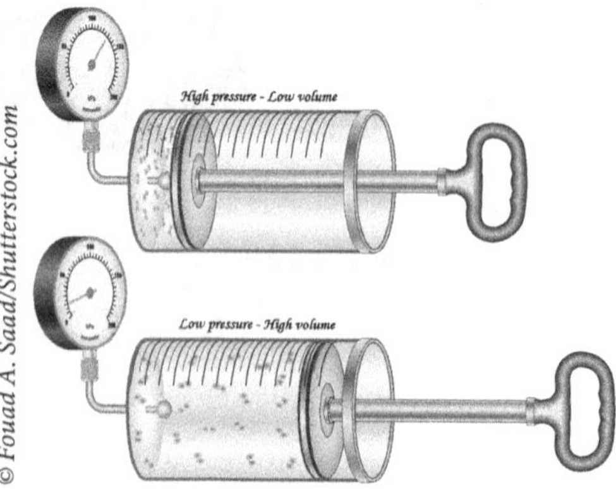

© Fouad A. Saad/Shutterstock.com

CHARLES'S LAW

Jacques Charles (1746–1823) isolated boron and studied gases. A balloonist, he determined that at constant pressure, the volume of a fixed mass of a given gas is directly proportional to the Kelvin temperature.

At constant pressure: $\frac{V_1}{T_1} = \frac{V_2}{T_2}$

© Neveshkin Nikolay/Shutterstock.com

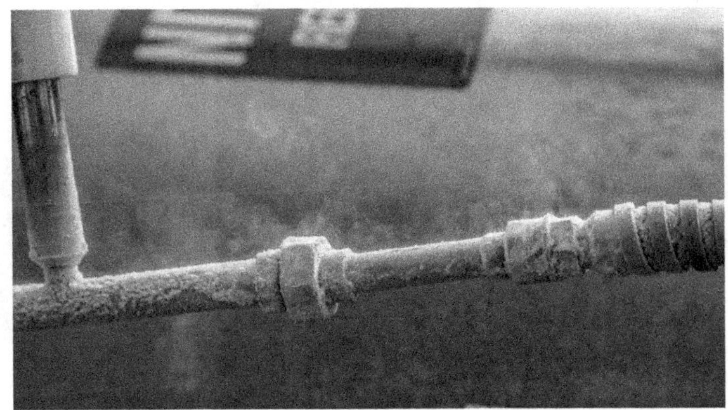

© Asmus/Shutterstock.com

GAY-LUSSAC'S LAW

Gay-Lussac (1778–1850) discovered that at constant volume, the pressure of a fixed mass of a given gas is directly proportional to the Kelvin temperature.

At constant volume: $\frac{P_1}{T_1} = \frac{P_2}{T_2}$

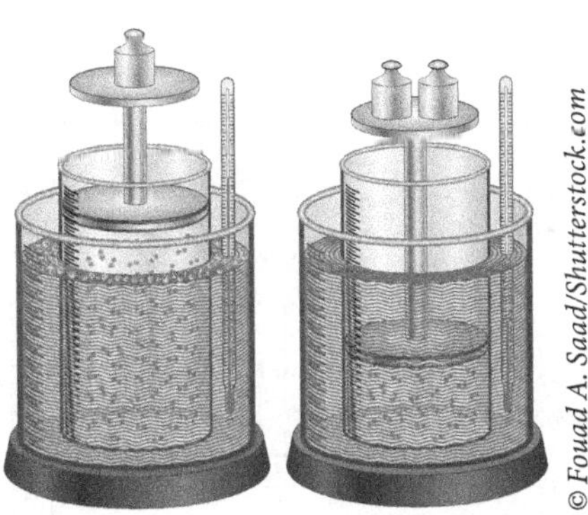

© Fouad A. Saad/Shutterstock.com

AVOGADRO'S LAW

Avogadro's law states that at constant temperature and pressure equal volumes of all gases contain the same number of moles.

At constant temperature and pressure: $\frac{V_1}{n_1} = \frac{V_2}{n_2}$

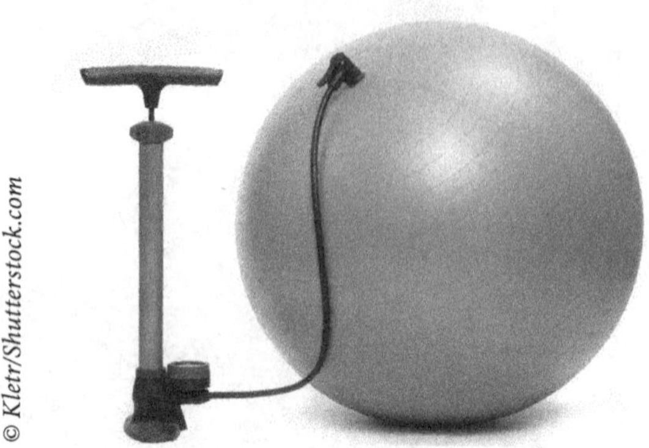
© Kletr/Shutterstock.com

COMBINED GAS LAW

The combined gas law states that at a fixed amount of gas, the pressure and volume of a gas is directly proportional to the temperature.

Therefore, nothing held constant: $\frac{P_1V_1}{n_1T_1} = \frac{P_2V_2}{n_2T_2}$

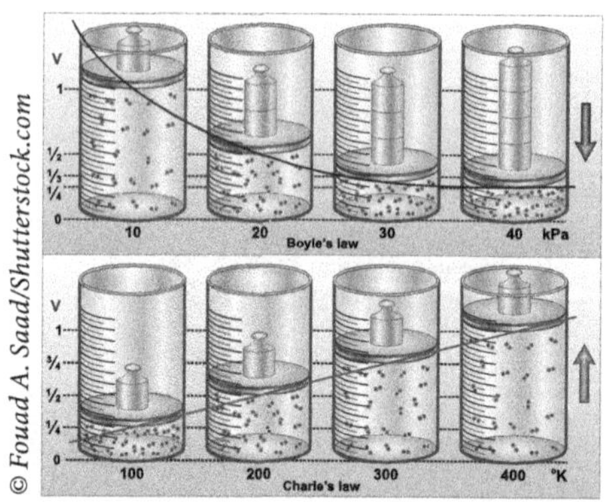

© Fouad A. Saad/Shutterstock.com

For example, anesthetic gas is normally given to a patient when the room temperature is 20.0°C and the patient's temperature is 37°C. What would this temperature change do to 1.60 L of gas if the pressure and the mass stay constant?

Pressure is constant

$$V_1 = 1.60L$$
$$T_1 = 20.0°C + 273.1 = 293.1K$$
$$V_2 = ?$$
$$T_2 = 37°C + 273 = 310.K$$

Equation: $V_2 = \dfrac{T_2V_2}{T_1}$

$$V_2 = \frac{(310.K)(1.60L)}{293.1K} = 1.69L$$

Example: After a sample of xenon with a volume of 0.532 L was heated from 22.0°C to 86.0°C, its volume changed to 587 mL and its pressure became 789 torr. What must have been its initial pressure in atmospheres?

$$P_1 = ?, P_2 = 789torr \times \frac{1atm}{760torr} = 1.04atm$$

$$V_1 = 0.532L, V_2 = 587mL \times \frac{1L}{1000mL} = 0.587L$$

$$T_1 = 22.0°C + 273.15 = 295.2K, T_2 = 86.0°C + 273.15 = 359.2K$$

$$P_1V_1T_2 = P_2V_2T_1$$

$$P_1 = \frac{P_2V_2T_1}{V_1T_2} = \frac{(1.04atm \times 0.587L \times 295.2K)}{(0.532L \times 359.2K)} = 0.943atm$$

THE IDEAL GAS LAW

The ideal gas law is a relationship between pressure, volume, temperature, and the number of moles of a gas. (All can be varied.)

PV = nRT (pronounced: puv-nert)

R = the ideal gas constant

$$R = 0.0821\frac{atmL}{molK}$$

$$R = 8.314\frac{J}{molK}$$

$$R = 62.36\frac{torrL}{molK}$$

This equation was derived by the following:

$$\frac{P_1V_1}{n_1T_1} = \frac{P_2V_2}{n_2T_2}$$

Set 1 mol of gas at STP

$$P_1 = 1atm$$
$$V_1 = 22.4L$$
$$n_1 = 1mol$$
$$T_1 = 0°C + 273 = 273K$$

$$\frac{P_1V_1}{n_1T_1} = \frac{(1atm)(22.4L)}{(1mol)(273K)} = 0.0821\frac{atm\,L}{mol\,K}$$

$$\text{Let } R = 0.0821\frac{atm\,L}{mol\,K}$$

$$R = \frac{PV}{nT}$$
$$PV = nRT$$

Example: Inflation of a bike tire.

When you increase the molecules into tire (n), the volume (V) increases slightly and the temperature (T) increases. When the volume cannot increase anymore, the tire becomes firm because the pressure increased (P).

Example: Oxygen sample.

A sample of oxygen at 24°C at 745 torr has a volume of 455 mL.

How many grams of O_2 were in the sample?

$$P = 745torr \times \frac{1atm}{760.torr} = 0.980atm$$

$$T = 24°C + 273 = 297K$$

$$V = 455mL \times \frac{1mL}{1000mL} = 0.455L$$

$$\text{Equation: } n = \frac{PV}{RT}$$

$$n = \frac{(0.980atm)(0.455L)}{(0.0821\frac{atm\,L}{molK})(297K)} = 0.0183molO_2$$

$$0.0183molO_2 \times \frac{32.0\,gO_2}{1molO_2} = 0.585\,gO_2$$

MOLECULAR MASS CALCULATIONS

Conversion:

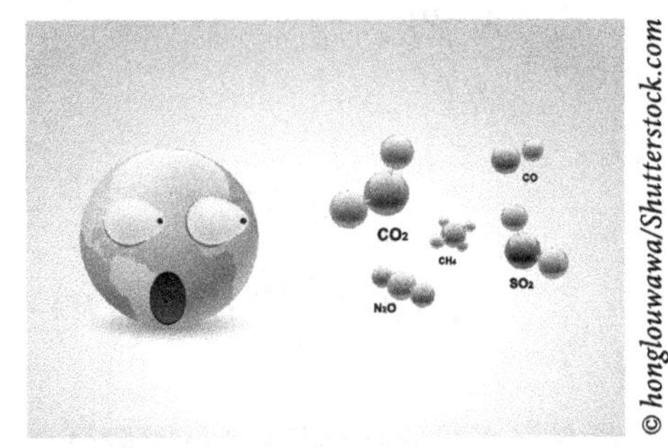

© honglouwawa/Shutterstock.com

$$PV = nRT$$

$$\text{If moles}(n) = \frac{\text{mass in grams }(m)}{\text{molar mass in g/mol }(M_m)}$$

$$PV = \frac{m}{M_m}RT$$

Rearrange equation to get:

$$M_m = \frac{mRT}{PV}$$

Example: Gas sample.

A sample of gas is collected in a 0.220 L gas bulb until its pressure reached is 0.757 atm at a temperature of 25.0°C. The sample's mass is 0.299 grams. What is the molecular mass of the gas?

$$M_m = \frac{mRT}{PV}$$

$$M_m = \frac{(0.299g)\left(0.0821\frac{atm\,L}{mol\,K}\right)(298.1K)}{(0.757atm)(0.220L)} = 43.9\,{}^{g}\!/_{mol}$$

DENSITY CALCULATIONS

Conversion:

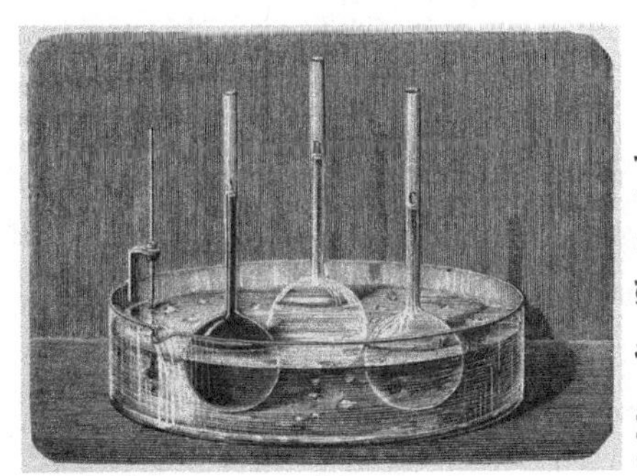

© Marzolino/Shutterstock.com

$$M_m = \frac{mRT}{PV}$$

$$\text{If density }(d) = \frac{\text{mass }(m)}{\text{volume }(V)}$$

and $m = dV$

$$M_m = \frac{dVRT}{PV} = \frac{dRT}{P}$$

Rearrange equation to get:

$$d = \frac{M_m P}{RT}$$

What density (in g/L) does oxygen have at 24.0°C and 742 torr?

$$d = \frac{M_m P}{RT}$$

$$d = \frac{\left(32.0\frac{g}{mol}\right)\left(0.976 atm\right)}{\left(0.0821\frac{atm L}{mol K}\right)\left(297.1 K\right)} = 1.28 \, {}^{g}\!/\!_{L}$$

GAS STOICHIOMETRY

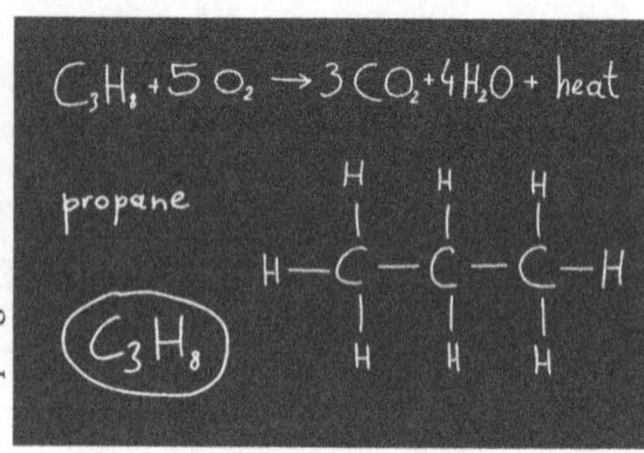

© Tupungato/Shutterstock.com

Gas stoichiometry involved the following steps:

Gas A Data → moles A → moles B → mass of B (or what is asked of B)

Example:

50.0 grams of CaC_2 is converted to C_2H_2.

$$CaC_2(s) + 2H_2O(l) \rightarrow Ca(OH)_2(s) + C_2H_2(g)$$

1. What is the theoretical yield of C_2H_2 in moles?

$$50.0 g \times \frac{1 mol CaC_2}{64.1 g} = 0.780 mol CaC_2$$

$$0.780 mol CaC_2 \times \frac{1 mol C_2H_2}{1 mol CaC_2} = 0.780 mol C_2H_2$$

2. In liters (at 24°C and 745 torr)?

$$V = \frac{nRT}{P} = \frac{\left(0.780 mol\right)\left(0.0821\frac{atm L}{mol K}\right)\left(297 K\right)}{0.980 atm} = 19.4 L C_2H_2$$

Look what happens when you compare volumes of gases in the same reaction, at STP.

Calculate the volume of SO_3 produced when 4.56 liters of SO_2 reacts with oxygen at STP.

Equation: $2SO_2(g) + O_2(g) \rightarrow 2SO_3(g)$

Doing Stoichiometry:

At STP: $4.56 L SO_2 \times \frac{1 mol SO_2}{22.4 L SO_2} = 0.204 mol SO_2$

Bridge Step: $0.204 mol SO_2 \times \frac{2 mol SO_3}{2 mol SO_2} = 0.204 mol SO_3$

At STP: $0.204 mol SO_3 \times \frac{22.4 L SO_3}{1 mol SO_3} = 4.56 L SO_3$

Since 1 mol of any gas = 22.4 liters at STP, the steps could be:

Bridge Step: $4.56 L SO_2 \times \frac{2 L SO_3}{2 L SO_2} = 4.56 L SO_3$

CHEMISTRY UNIT 8 PRACTICE PROBLEMS

Convert Units for Gas Laws

1. Convert 744 torr to atmospheres and to kPa.

2. How many torr is equivalent to 1.25 atm?

3. What is 20.5°F to Kelvin?

4. Convert 357 mL to liters.

5. How many liters are in 1.125 grams of carbon dioxide gas at STP?

Combined Gas Laws

6. If an amount of gas at 0.978 atm and 0.456 L is compressed at a constant temperature to a volume of 0.250 L, what is the pressure?

7. What volume would a hydrogen balloon be at 725 torr, if the temperature is constant and it had a volume of 1.527 L at 759 torr?

8. An aerosol can has a pressure of 1.30 atm at 25°C. What is the pressure at 200°C? At constant volume?

9. A diver set a tanks pressure to 770.25 torr at 20°C, in the water the pressure changes to 745.6 torr what is the temperature of the tank in the water? At constant volume?

10. Given a 526 mL sample of a gas at 22.7°C and 748.9 torr, what is the volume if the temperature and pressure are at STP?

11. A gas has a volume of 0.00400 mL at 25.5 atm and 25°C. What is the pressure if the volume of the gas changed to 8.00 mL and the temperature changed to 55°C?

12. A gas pressure at 781 torr at 20°C in a 456 mL tank has a pressure change to 745.6 torr. What is the temperature of the tank in the water at constant volume?

13. A hot air balloon has a volume of 1600 L at 180°C. What is the volume if the gas is heated to 220°C?

14. At 293.6K a gas has a volume of 356.3 mL. What is the temperature if the volume changes to 405.8 mL?

GAS LAWS: IDEAL GAS

1. If I have 4 moles of a gas at a pressure of 5.6 atm and a volume of 12 liters, what is the temperature?

2. If I have 21 moles of gas held at a pressure of 3800 torr and a temperature of 627°C , what is the volume of the gas?

3. If I have 54.3 grams of nitrogen held at a pressure of 5 atm and in a container with a volume of 50 liters, what is the temperature (in Celcius) of the gas?

4. Calculate the molecular mass of 4.5 grams of a gas at a volume of 345 mL and a pressure of 45.8 psi at 25°C.

5. A gas exerts a pressure of 0.892 atm in a 523 mL container at 15°C. The density of the gas is 1.22 g/L. What is the molecular mass of the gas?

6. Determine the density of helium at STP.

ASSOCIATION WITH GASES: PRESSURE, VOLUME, TEMPERATURE, AND MOLES

1. Perform the following conversions:
   a. 453 kPa = _____ atm
   b. 281.3 cm^3 = _____ L
   c. 61°C = _____ °F = _____ K
   d. 92.1 g Cl_2 = _____ molCl_2

2. Calculate the following volume at STP for each:
   a. 0.45 mol N_2 = _____ L N_2
   b. 2.3 × 10^4 mol Ar = _____ mL Ar
   c. 92.1 g Cl_2 = _____ L Cl_2

3. The pressure of nitrogen in a cylinder is 2.00 × 10^3 psi. What is the pressure in atmospheres?

4. The vapor pressure of mercury is 0.0012 torr at 20°C. What is this pressure in atmospheres?

COMBINED GAS LAW

1. Use the gas laws to calculate the following:
   a. 1.2 L of carbon disulfide is heated from 15°C to 80°C at constant pressure. What is the new volume?

   b. 155 mL of freon is taken from 810 torr to standard pressure. What is the new volume at constant temperature?

   c. 4.12 L of oxygen is at 0.977 atmospheres and 15.2°C. What is the volume at 1.04 atmospheres and 42.0°C?

2. Consider a cylinder fitted with a movable piston. The initial pressure inside the cylinder is P_i and the initial volume is V_i. What is the new pressure in the system when the piston decreases the volume of the cylinder by half?
   a. 2 V_iP_i
   b. (1/4) P_i
   c. P_i^2
   d. 2 P_i
   e. (1/2) P_i

3. The pressure of a combustible mixture in a cylinder of a motorcycle engine is 0.980 atm when the volume is 246 mL. The piston decreases the volume to 24.1 mL. What is the pressure (atm) at that point assuming no change in temperature occurs?

4. What would be the volume (mL) of a sample of ethane (C_2H_6) at 467 K and 1.2 atm if it occupied 405 mL at 298 K and 1.2 atm?

5. A balloon is filled with air and has a volume of 3.25 L at 30°C. The balloon is placed in a freezer at –10°C. What is the volume of the balloon at this temperature?

6. A sample of a gas at –91°C and 1 atm occupies 2.0 L. What volume, L, will the gas occupy at 0°C at the same pressure?

7. A spray can is used until only the propellant gas remains at a pressure of 1.1 atm at 23°C. If the can were thrown into a fire at 475°C, what would be the pressure (atm) in the hot can?

8. Gas evolved in the fermentation of sugar in wine making occupies a volume of 0.75 L at 20°C at 720 mm Hg. What volume (L) would the gas occupy at 39°C and 1.00 atm?

9. A gas occupies 1.0 L at 27°C and 0.50 atm. At what temperature (°C) will the gas occupy at 0.50 L at 1.0 atm?

IDEAL GAS LAW

1. Use the ideal gas law to calculate the following:
   a. What pressure will be exerted by 1.42 moles of butane, C_4H_{10}, in a 10.0 L cylinder at 75°C?

   b. A sample of oxygen is collected in a laboratory experiment. Find its mass if the volume is 618 mL at 790 torr and 15°C.

2. A 25L cylinder contains 128 g of nitrogen gas at 10°C. How many grams of nitrogen must be added to increase the pressure to 5.00 atm assuming ideal gas behavior?

3. A balloon is filled with 48.3 g of helium at 31°C and 2.12 atm. What is the volume of the balloon in liters?

4. A 40.0 gram sample of helium is introduced into a 2.24 L cylinder which is heated until the pressure is 200 atm. What is the temperature (in °C) of the gas?

5. An incandescent light bulb with a volume of 125 cm^3 contains 2.5×10^{-3} moles of argon. What is the pressure of argon (atm) at 25°C?

6. The pressure in a 2.0 L container is 1.5×10^{-4} torr at 1115 K. How many moles are in the container?

7. The Goodyear blimp has 5.12×10^6 liters of helium at 25°C and 1.00 atm. What mass, g, of helium is in the blimp?

8. A 725 gram sample of neon is introduced into a 4.5 L cylinder which is then heated until the pressure is 225 atm. What is the temperature (°C) of the gas?

9. What volume is occupied by 0.0100 mole of carbon monoxide at 9°C and 0.973 atm?

10. What is the temperature of 0.0250 mole of hydrogen gas if it occupies 665 mL at 715 torr?

MOLECULAR MASS AND DENSITY

1. Use the molecular mass and density formulas to calculate the following:
   a. 4.22 grams of a gas has a volume of 2.32 liters and a pressure of 0.211 atm. If the temperature is 25.0°C, calculate the molar mass.

   b. What is the density in g/L of BF_3 at STP?

   c. A gaseous compound was found to have a density of 5.60 g/L at 23.0°C ant 750 torr. Calculate the molar mass.

2. What is the density (g/L) of nitrogen at STP?

3. What is the density in g/L of BrF_3 at 425 torr and 77°C?

4. Calculate the density of neon at 32°C and 0.676 atm.

5. Calculate the density of nitrogen dioxide at STP.

GAS STOICHIOMETRY

1. Using stoichiometry, calculate the following:
   a. What volume of oxygen gas at STP is produced by the decomposition of 100.0 grams of sodium nitrate?

   $2NaNO_4 \rightarrow 2NaNO_3 + O_2$

   b. How many liters of CO_2 are produced at 25.0°C at 760. torr by burning 24.0 grams of CH_4 in oxygen?

   $CH_4 + 2O_2 \rightarrow CO_2 + 2H_2O$

2. What volume of hydrogen (in L) at STP would be required to react with 0.100 mole of nitrogen to form ammonia?

   $N_2(g) + 3H_2(g) \rightarrow 2NH_3(g)$

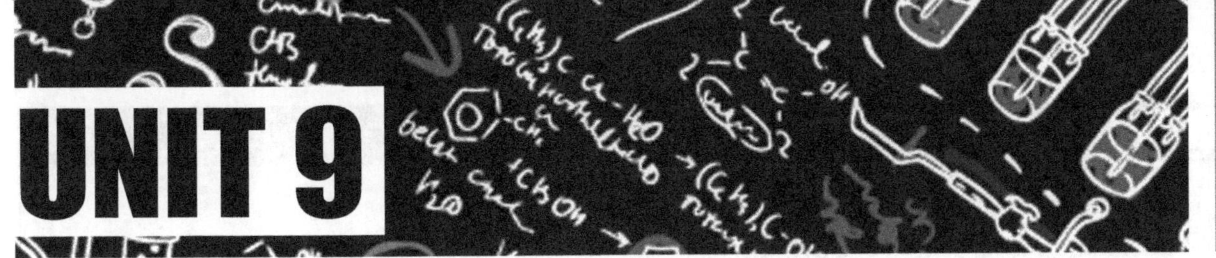

FINDING THE SOLUTIONS (SOLUTIONS AND ACID-BASE REACTIONS)

The Characteristics of a Solution

Ionic Compounds Dissolved in Water

Molecular Nonacid Compounds Dissolved in Water

Molecular Acid Compounds Dissolved in Water

Solution Terminology: Solubility, Unsaturated, Saturated, Supersaturated, Crystallization

Hydrated Compounds

Solution Concentration: Percent by Mass

Solution Concentration: Molarity

Dilution of Concentration Solutions

Titration Using Molarity

Acid-Base Titration

Calculations Involving Neutralization Reactions

Acids and Bases

Naming Acid Molecules
Naming Base Molecules

Neutralization Reactions

Arrhenius Theory of Acids and Bases

Brønsted-Lowry Theory of Acids and Bases

Identifying Acids, Bases, Conjugate Acids, and Conjugate Bases

Strong and Weak Acids

Strong and Weak Bases

The Water Equilibrium & pH and pOH

OBJECTIVES FOR UNIT 9

- To define a solution and understand its terminology.
- To understand the formation of a hydrate.
- To perform concentration calculations of percentage by mass and molarity.
- To perform calculations based on diluting solutions.
- To understand a titration and the calculations involved.
- To define the Arrhenius Theory and Brønsted-Lowry theory of acids and bases.
- To classify strong and weak acids and bases.
- To identify acids, bases, conjugate acids, and conjugate bases in a chemical reaction.
- To understand the pH and pOH scale for identifying the concentration of hydrogen ions and hydroxide ions in solution.

THE CHARACTERISTICS OF A SOLUTION

Aqueous solutions are solutions that involve a substance being dissolved (solute) and water doing the dissolving (solvent). Water is considered the "universal solvent" because it can dissolve many different substances.

IONIC COMPOUNDS DISSOLVED IN WATER

Ionic compounds "break apart" into their separate ions when dissolved in water. These ions are:

Let's review what happens when ions are put into solution.

Take NaCl

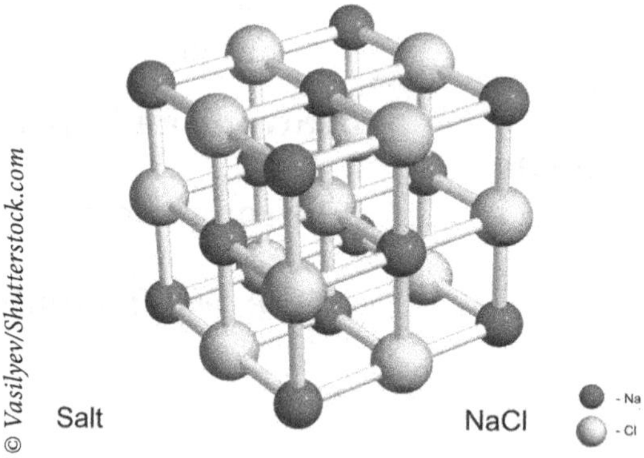

© Vasilyev/Shutterstock.com

In the solid state, NaCl is a crystal lattice.

Dissolving NaCl in water breaks the ions apart:

These are **hydrated ions** because they are surrounded by water molecules.

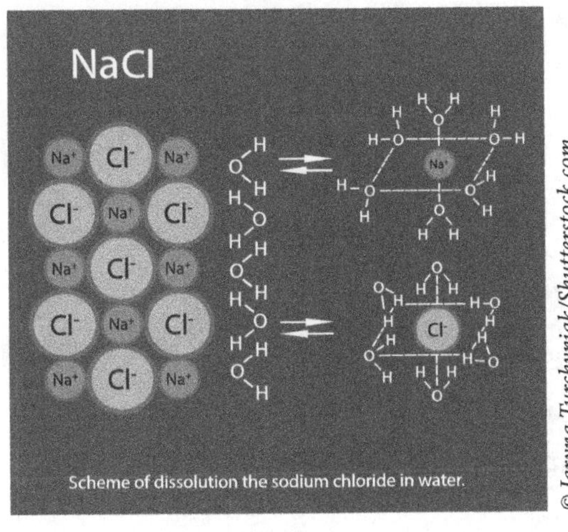

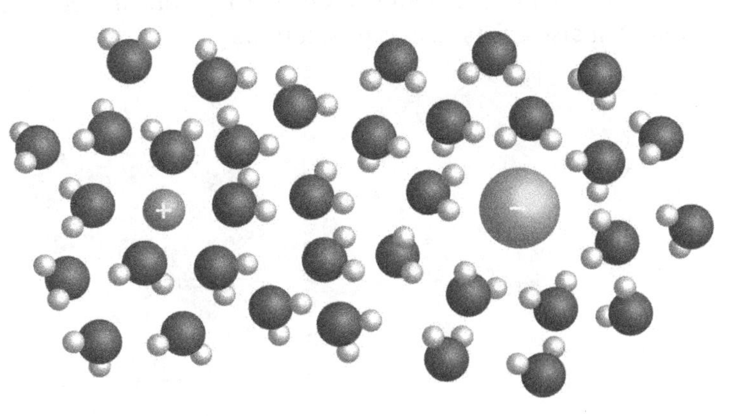

Eventually, all the ions are in solution.

MOLECULAR NONACID COMPOUNDS DISSOLVED IN WATER

Because water is a polar compound and "like dissolves like," water can dissolve many molecular compounds that are also polar. Two situations occur when water dissolves molecular compounds.

Solvated compounds: The molecular compounds are "surrounded" by water molecules, which keep them in solution (much like in the case of ionic compounds).

Example: A sugar molecule "dissolved" in water.

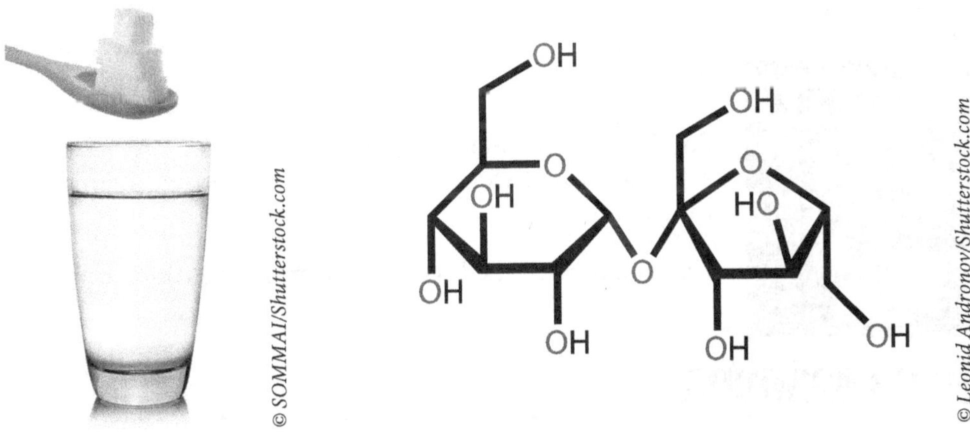

MOLECULAR ACID COMPOUNDS DISSOLVED IN WATER

Ionization: The molecular compounds actually get "pulled apart" into separate ions by the strong charges on the water molecules. The shared bonds actually get broken.

Example: Let's examine the reaction of HCl (aq) and NaOH (aq) in details.

$$HCl(aq) + NaOH(aq) \rightarrow NaCl(aq) + H_2O(l)$$

In aqueous solution, HCl and NaOH exist as ions in solution. The salt, NaCl, that forms is soluble in water; therefore, it also exists as ions in solution.

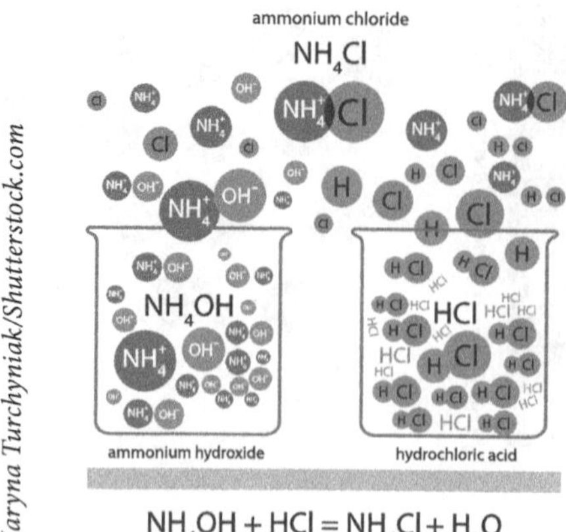

SOLUTION TERMINOLOGY: SOLUBILITY, UNSATURATED, SATURATED, SUPERSATURATED, CRYSTALLIZATION

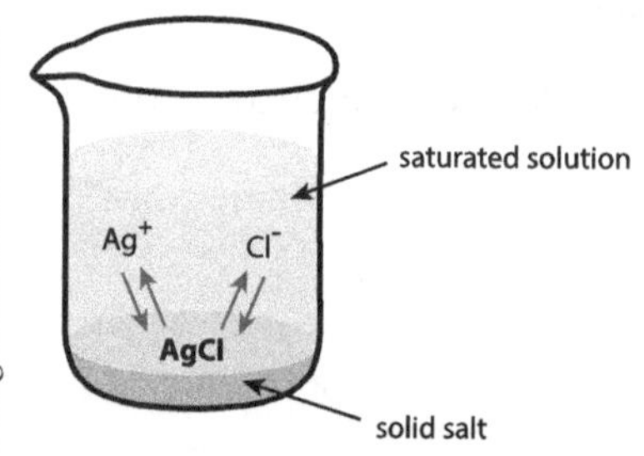

Solubility of a solute is the maximum amount of a solute that can dissolve in a solvent.

Concentration is used to describe the amount of solute dissolved in a solvent.

Unsaturated solutions contain less than the maximum amount of solute is dissolved in a solvent.

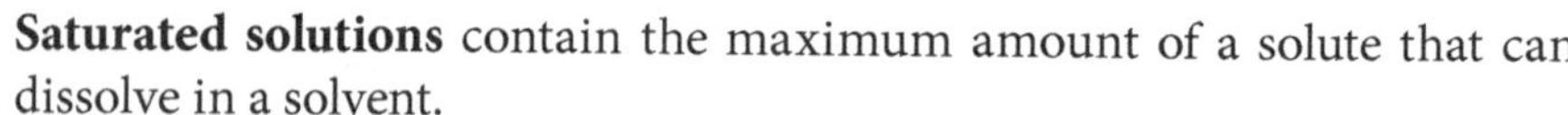

Saturated solutions contain the maximum amount of a solute that can dissolve in a solvent.

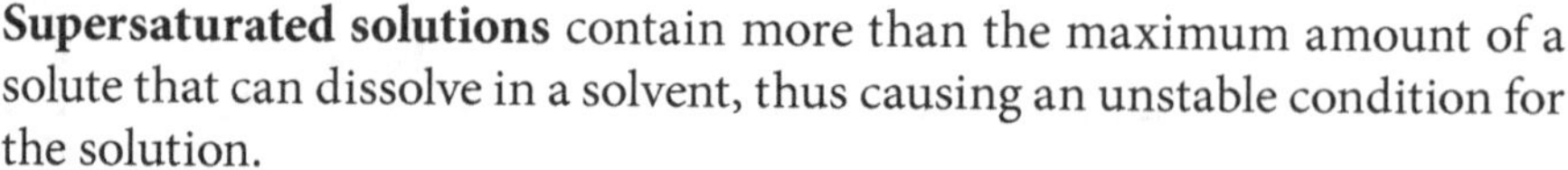

Supersaturated solutions contain more than the maximum amount of a solute that can dissolve in a solvent, thus causing an unstable condition for the solution.

Under supersaturated conditions, **crystallization** occurs very readily.

HYDRATED COMPOUNDS

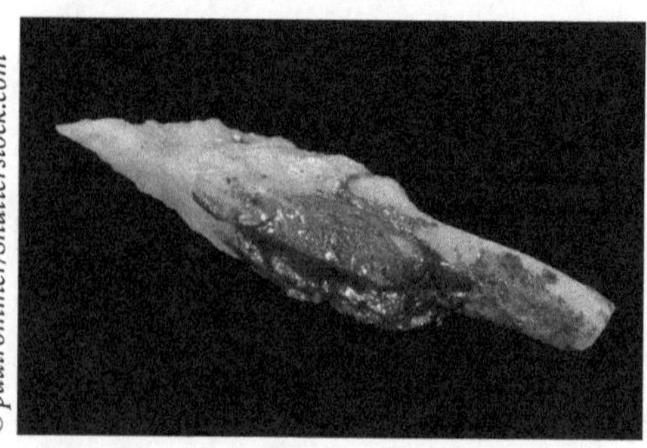

Recall that hydrates are ionic compounds with a formula including water molecules attached. Often hydrates are formed when some compounds are exposed to water vapor in our atmosphere. These compounds "attract" water molecules by its cations.

Example:

Fe_2O_3 is a hematite mineral (compound for rust) which has a rhombohedral crystal (a hexagonal) system. When it combines with water it can form the hydrate $Fe_2O_3•3H_2O$, which is a limonite mineral—an amorphous mineral-like substance.

The picture of a hydrate compound: Copper(II) sulfate pentahydrate is used to kill algae in pool and reservoir water, and is used as a copper source in animal feeds.

SOLUTION CONCENTRATION: PERCENT BY MASS

Percent by mass relates the mass of solute as the total mass of the solution.

Equation: $\% \text{ by Mass} = \dfrac{\text{mass of solute}}{\text{mass of solution}} \times 100$

Example: Sodium phosphates is an oral solution available in the United States as an over-the-counter preparation indicated "for the relief of occasional constipation." It is professionally labeled "for use as part of a bowel cleansing regimen in preparing the patient for surgery or preparing the colon for x-ray or endoscopic examination."

Calculate the mass percent of Na_3PO_4 if 5,655 grams of Na_3PO_4 is dissolved in 8,000 grams of water.

$$\frac{5{,}655\,g}{5{,}655\,g + 8{,}000\,g} \times 100 = 41.4\%$$

SOLUTION CONCENTRATION: MOLARITY

When you make a Kool-Aid® drink, you have made a solution. If the Kool-Aid® is too sweet, add more water. If the Kool-Aid® is not sweet enough, add more sugar.

Thankfully, the Kool-Aid® package lets us know exactly how much sugar to add to a certain amount of water. This "recipe" on the package actually reports the concentration of the Kool-Aid® solution.

Molarity is another type of concentration, known as the molar concentration. **Molarity** is the moles of solute divided by the liters of solution.

First: 1.00 cup $C_{12}H_{22}O_{11}$ = 192 grams $C_{12}H_{22}O_{11}$

2.00 quarts of solution = 1.89 liters solution

$$192\,g\ C_{12}H_{22}O_{11} \times \frac{1\,mole}{342.30\,g} = 0.561\,mole$$

$$M = \frac{0.561\,mol}{1.89\,L} = 0.297\,M\ C_{12}H_{22}O_{11}$$

What does molarity tell us in terms of preparing a solution?

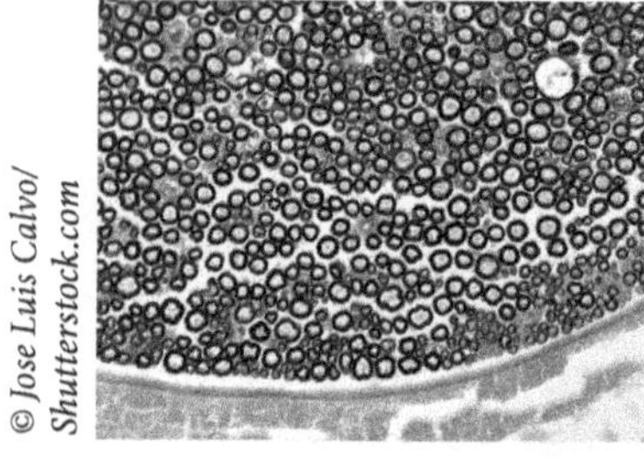
© Jose Luis Calvo/ Shutterstock.com

Osmium tetroxide, OsO_4, is a very poisonous pale, yellow solid that is used as a fixing and staining agent for cell and tissue studies. It is one of the best methods for revealing lipids. It binds at double bonds of unsaturated lipids and imparts a dense brownish or black color. In the picture to the left, you see the effects of this stain; the lipid-rich myelin sheath of nerve fibers is heavily colored.

How many grams of OsO_4 are needed to prepare 500. mL of a 0.00034 M OsO_4 solution?

$$0.500L \times \frac{0.00034\,mol}{1L} = 0.000170\text{mol}$$

$$0.000170\text{mol} \times \frac{254.23\,g}{1\,mol} = 0.0432\text{g OsO}_4$$

© rdonar/Shutterstock.com

Example: Barium bromate is a chemical used for coating wire.

$Ba(BrO_3)_2$ dissolved in water to make up one liter of solution. The ion concentration is as follows:

$Ba(BrO_3)_2 \rightarrow Ba^{2+}\,(aq) + 2\,BrO_3^{1-}(aq)$

DILUTION OF CONCENTRATION SOLUTIONS

Equation:

Example: 20. mL of 5.6 *M* NaOH was diluted to 50. mL. What is the new concentration of NaOH?

$$M_1V_1 = M_2V_2$$

$$M_2 = \frac{M_1V_1}{V_2} = \frac{(20.ml \times 5.6M)}{50.mL} = 2.2M\ NaOH$$

TITRATION USING MOLARITY

ACID-BASE TITRATION

Titration is an analytical procedure that allows us to measure the amount of one solution (a base) needed to react exactly with the contents of another solution (an acid). Such analyses, which involve the measurements of volumes of solutions of reactants, are called volumetric analyses.

Some important terminology:

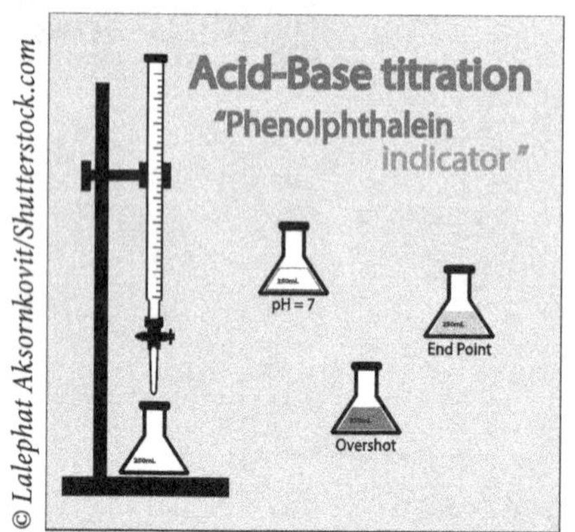

© Lalephat Aksornkovit/Shutterstock.com

1. Buret: A long tube fitted at one end with a valve (stopcock) that precisely measures milliliter amounts.
2. Titrant: The solution in the buret (usually the base).
3. Indicator: A dye added to the solution in the receiving vessel, it undergoes color change when reaction is over.
4. Phenolphthalein: colorless in acid, pink in base.
5. Equivalence point: When the indicator has suggested the reaction to be over.
6. Standardization: The procedure by which the concentration of an analytical reagent is determined.

CALCULATIONS INVOLVING NEUTRALIZATION REACTIONS

Calculations involving neutralization reactions are usually carried out after performing the titrations. Follow the steps outlined below to solve this type of problems.

Calculate the volume of a 0.200 M KOH solution that is needed to neutralize 25.00 mL of a 0.115 M HCl solution.

Step 1: Write a balanced equation for the neutralization reaction and determine the mole ratio of the acid to base.

$$1\ KOH(aq) + 1\ HCl(aq)\ 1\ KCl(aq) + 1\ H_2O(l)$$

The mole ratio of KOH to HCl is 1 mol KOH to 1 mol HCl

Step 2: List the volume and molarity for the HCl and the KOH solutions.

	HCl	KOH
Volume	25.00 mL	?
Molarity	0.115 M	0.200 M

Step 3: Determine the availability of the number of moles of HCl that is available in the titration.

There is a specific number of moles in 25.00 mL of a 0.115 M HCl.

$$0.02500L \times 0.115mol/L = 0.00288mol\ HCl$$

Step 4: Apply the mole ratio determined in step 1 to determine the moles of base that is needed to neutralize the available acid.

0.00288mol HCl = 0.00288mol KOH because the mole ratio is 1 mol HCl to 1 mol KOH

Step 5: Divide the moles of base by the given concentration of KOH.

The concentration of KOH solution is 0.200 M

0.00288 moles of KOH are needed to neutralize the available HCl

Therefore, rearranging the Molarity equation:

$$Volume = \frac{moles}{Molarity} = \frac{0.00288mol}{0.200\frac{mol}{L}} = 0.0144L\ or\ 14.4mL$$

ACIDS AND BASES

Let's review what we have already learned about acids and bases.

Acids: citrus fruits, aspirin, soda pop, vinegar, and vitamin C

Bases: baking soda, detergents, ammonia cleaners, antacids, and soap

NAMING ACID MOLECULES

Oxoacid molecules have oxygen present.

Compound	Anion	Acid Name
HClO	hypochlorite, ClO^-	hypochlorous acid
$HClO_2$	chlorite, ClO_2^-	chlorous acid
$HClO_3$	chlorate, ClO_3^-	chloric acid
$HClO_4$	perchlorate, ClO_4^-	perchloric acid
H_2CO_3	carbonate, CO_3^{2-}	carbonic acid

If the polyatomic ion that the acid is derived from ends in:

-ite it changes to *-ous* acid

-ate it changes to *-ic* acid

Nonoxoacid molecules have no oxygen present.

Stomach acid is the strong acid HCl. This acid does not contain oxygen so we refer to it as a nonoxoacid.

Naming nonoxoacids involve using the prefix: *hydro-* and the ending *-ic* acid.

Compound	Anion	Acid Name
HCl	chloride, Cl^-	hydrochloric acid
HF	fluoride, F^-	hydrofluoric acid
H_2S	sulfide, S^{2-}	hydrosulfuric acid

Note: sulfide does not change to hydrosulfuric acid; it is properly written hydrosulfuric acid.

NAMING BASE MOLECULES

Base molecules follows the rules for naming compounds.

Example: Lye solutions are chemically known as NaOH (sodium hydroxide).

NEUTRALIZATION REACTIONS

Neutralization reactions occur when an acid reacts with a base to produce a salt and water.

Acid + Base → Salt + Water

Let's examine the reaction of HCl (aq) and NaOH (aq) in detail.

$$HCl(aq) + NaOH(aq) \rightarrow NaCl(aq) + H_2O(l)$$

We learned that, in aqueous solution, HCl and NaOH exist as ions in solution. The salt, NaCl, that forms is soluble in water; therefore, it also exists as ions in solution. Therefore, to describe the same reaction more precisely, we write the following reaction revealing the ions that exist in the solution.

$$H^+(aq) + Cl^-(aq) + Na^+(aq) + OH^-(aq) \rightarrow Na^+(aq) + Cl^-(aq) + H_2O(l)$$

Careful inspection of the beakers will reveal that the ions that have undergone reaction are the H^+ and OH^- ions. The Na^+ ions and Cl^- ions remain unchanged before and after the reaction. The Na^+ ions and the Cl^- ions are called spectator ions.

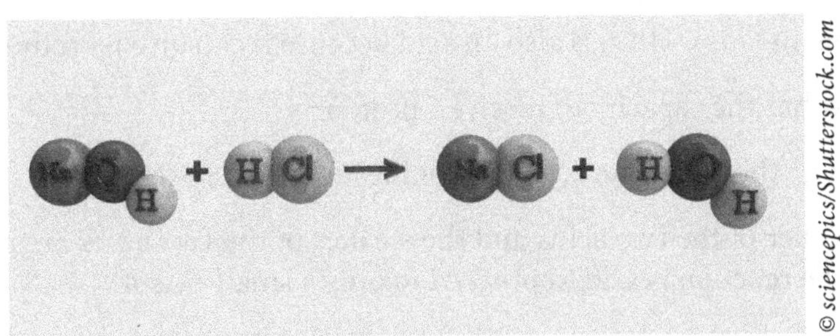

© sciencepics/Shutterstock.com

If we leave out the spectator ions and write the equation involving those ions that participated in the reaction, we write the net ionic equation of neutralization.

$$H^+(aq) + OH^-(aq) \rightarrow H_2O(l)$$

ARRHENIUS THEORY OF ACIDS AND BASES

Svante August Arrhenius (1859–1927), winner of the 1903 Noble Prize in Chemistry.

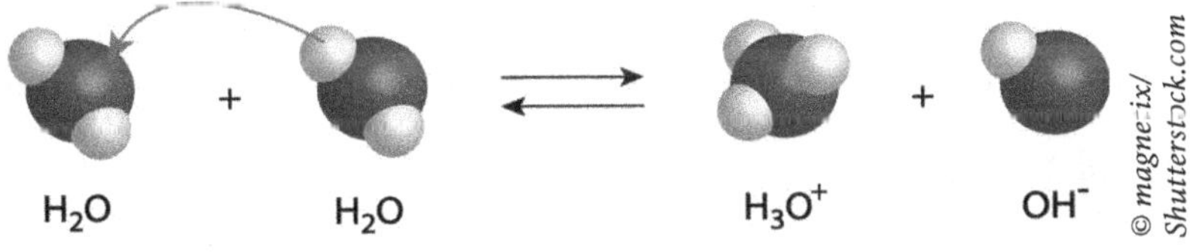

© magnetix/Shutterstock.com

THE BRØNSTED-LOWRY THEORY OF ACIDS AND BASES

In 1923, within several months of each other, Johannes Nicolaus Brønsted (Denmark) and Thomas Martin Lowry (England) published essentially the same theory about how acids and bases behave. Since they came to their conclusions independently of each other, both names have been used for the theory name.

Note: For the arrows that indicate equilibrium, the arrow will always show favoritism towards the weaker acid and weaker base.

Water can behave as both and acid and a base, this behavior is called amphoteric.

IDENTIFYING ACIDS, BASES, CONJUGATE ACIDS, AND CONJUGATE BASES

Example:

$$HNO_3 + H_2O \rightleftharpoons H_3O^+ + NO_3^-$$

HNO_3 is an acid, because it has a proton available to be transferred.

H_2O is a base, since it gets the proton that the acid lost.

Now, here comes an interesting idea: H_3O^+ is also an acid because it can give a proton.

NO_3^- is also a base, since it has the capacity to receive a proton.

Thus, we call H_3O^+ and NO_3^- the conjugate acid and conjugate base, respectively.

More Examples: If the weaker of the two acids and the weaker of the two bases are reactants (appear on the left side of the equation), the reaction is said to proceed to only a small extent:

$$HC_2H_3O_2 + H_2O \rightleftharpoons H_3O^+ + C_2H_3O_2^-$$
$$NH_3 + H_2O \rightleftharpoons NH_4^+ + OH^-$$

Identify the conjugate acid base pairs in each reaction.

STRONG AND WEAK ACIDS

Properties: Sour taste, produces H+ ions (or sometimes represented as H_3O^+ ions) when dissolved in water. Undergoes double replacement reactions with solid oxides, hydroxides, carbonates, and bicarbonates (illustrated later). Acids release a hydrogen ion into water (aqueous) solution. Acids neutralize bases. Acids corrode active metals. Acids turn blue litmus to red.

Note: The hydrogen ion that is produced in solution is also illustrated with the hydrodium ion H_3O^+

"Strong" acids ionize completely in water.

Example: HCl, stomach acid

$$HCl(aq) + H_2O(l) \rightarrow H_3O^+(aq) + Cl^-(aq)$$

Hydrochloric acid completely dissociates into H_3O^+ ions (or sometimes represented as H^+ ions) and Cl^- ions when dissolved in water. Therefore, HCl is a strong acid.

Common Strong Acids: HCl, HBr, HI, HNO_3, H_2SO_4*, $HClO_3$, $HClO_4$

"Weak" acids ionize partially in water.

Example: Vinegar is an acetic acid solution.

$$CH_3COOH(aq) + H_2O(l) \rightleftharpoons H_3O^+(aq) + CH_3COO^-(aq)$$

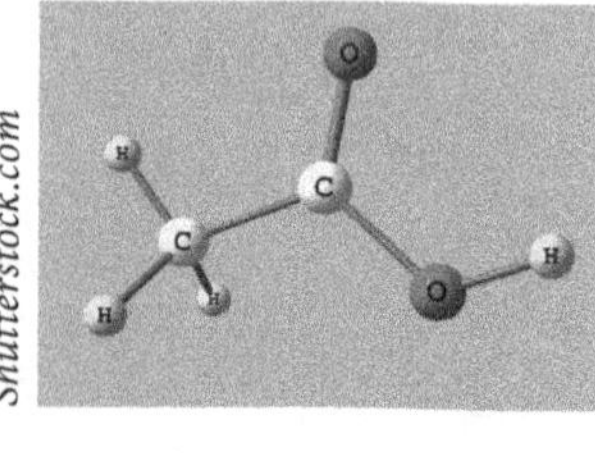

Acetic acid partially dissociates into H_3O^+ ions and CH_3COO^- ions when dissolved in water. Therefore, acetic acid is a weak acid.

Note: The $\rightleftharpoons$ indicates the limited ionization (or dissociation) of weak acids.

Common weak acids include HF, H_2S, HNO_2, $HClO_2$, HClO, H_2SO_3, H_3PO_4, H_2CO_3, H_3BO_3, CH_3COOH, HSO_4^-.

STRONG AND WEAK BASES

Properties: Base solutions have a slippery or soapy feel and taste bitter. They will react with acids to produce a salt. Bases release a hydroxide ion into water solution. Bases neutralize acids. Bases denature protein. Bases turn red litmus to blue.

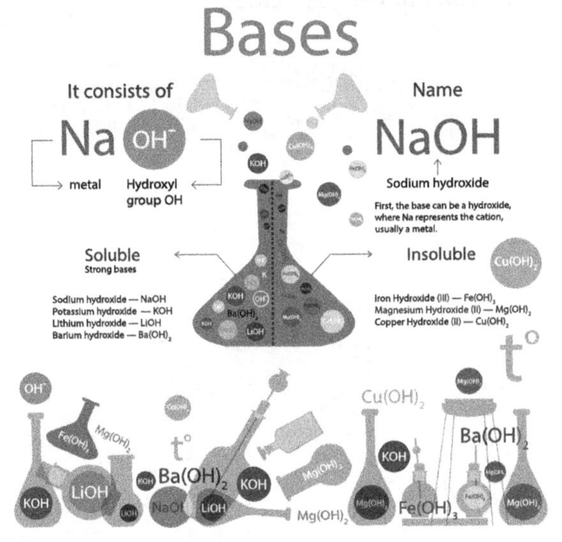

© Iaryna Turchyniak/Shutterstock.com

"Strong" bases ionize completely in water.

Example: Caustic Soda (sodium hydroxide), NaOH, is an important ingredient for making soaps.

$$NaOH(aq) + H_2O(l) \rightarrow Na^+(aq) + OH^-(aq)$$

Sodium hydroxide completely dissociates into Na^+ ions and Cl^- ions when dissolved in water. Therefore, NaOH is a strong base.

Common strong bases include LiOH, NaOH, KOH, RbOH, CsOH, $Ba(OH)_2$, $Ca(OH)_2$, $Sr(OH)_2$, $Mg(OH)_2$.

"Weak" bases ionize partially in water.

Example: Ammonia

NH_3 + H_2O → NH_4^+ + OH^-

© magnetix/Shutterstock.com

$$NH_3(aq) + H_2O(l) \rightleftharpoons NH_4^+(aq) + OH^-(aq)$$

Ammonia partially dissociates into NH_4^+ ions and OH^- ions when dissolved in water. Therefore, ammonia is a weak base.

Note: The $\rightleftharpoons$ indicates the limited ionization (or dissociation) of weak bases.

Common weak bases include NH_3, CH_3NH_2.

THE WATER EQUILIBRIUM AND pH AND pOH

pH is a measure of the $[H_3O^+]$ in solution.

Developing the concept of pH:

Water ionizes according to the equation: $H_2O_{(l)} \rightleftharpoons H^+_{(aq)} + OH^-_{(aq)}$

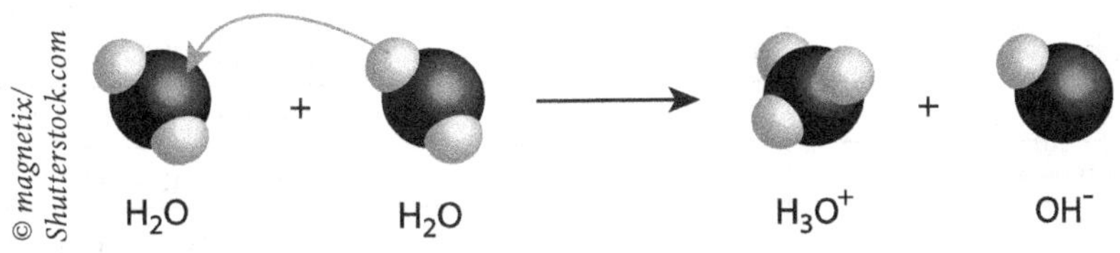

© magnetix/Shutterstock.com

Experimental evidence indicates that pure water contains 1×10^{-7} mole of both H^+ and OH^-

This leads to the pH Scale:

The PH Scale

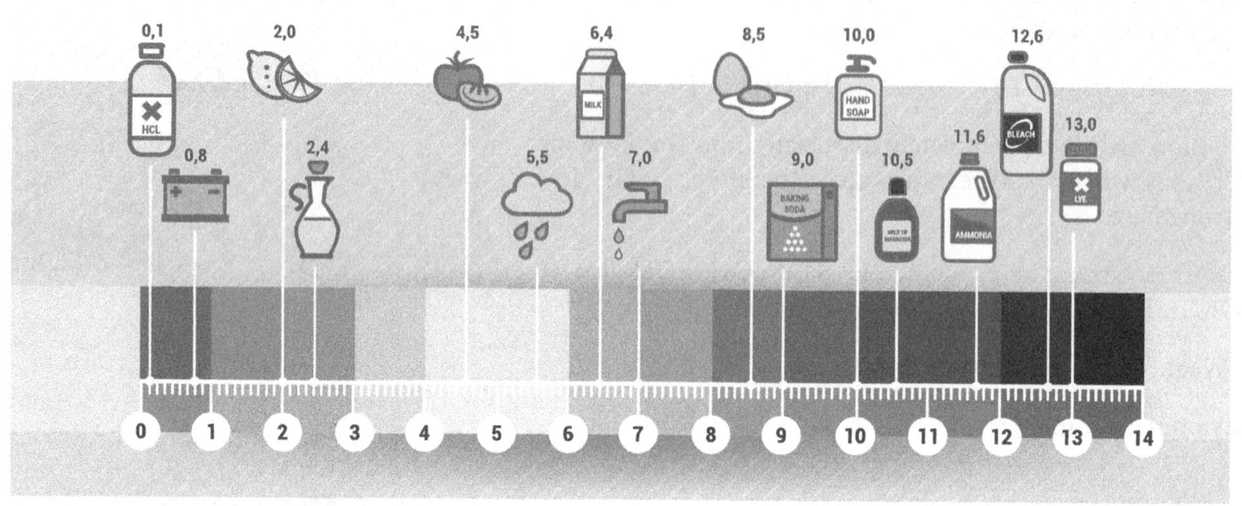

Acidic — **Neutral** — **Alkaline**

Lorem ipsum dolor sit amet

Consectetur adipisicing elit, sed do eiusmod tempor incididunt ut labore et dolore magna aliqua. Ut enim ad minim veniam, quis nostrud voluptate velit esse cillum dolore eu fugiat nulla pariatur. Excepteur sint occaecat cupidatat non proident, sunt in culpa qui officia deserunt mollit anim id est laborum. Lorem ipsum dolor sit amet, consectetur adipisicing elit, sed do eiusmod tempor incididunt ut labore et dolore magna aliqua.

Ut enim ad minim veniam

Quis nostrud exercitation ullamco laboris nisi ut aliquip ex ea commodo consequat. Duis aute irure dolor in reprehenderit in voedo consequat. Duis aute iesse cillum dolore eu fugiat nulla pariatur. Excepteur sint ocnon proident, sunt in culpa qui officia deserunt mollit anim id est laborum. Lorem ipsum dolor sit amet, consectetur adipisicing elit, sed do eiusmod tempor incididunt ut labore et dolore magna aliqua.

Duis aute irure dolor

In reprehenderit in voluptate velit esse cillum dolore eu fugiat nulla pariatur. Excepteur sint occaecat cupidatat non proident, sunt iullamprehenderit in voluptate velit esse cillum dolore eu fugiat nulla pariatur. Excepteur sint non proident, sunt in culpa qui officia deserunt mollit anim id est laborum. Lorem ipsum dolor sit amet, consectetur adipisicing elit, sed do eiusmod tempor incididunt ut labore et dolore magna aliqua.

Lorem ipsum dolor sit amet

Consectetur adipisicing elit, sed do eiusmod tempor incididunt ut labore et dolore magna aliqua. Ut enim ad minim veniam, quis nostrud exercitation ullamco laboris niln voluptate velit esse cillum dolore eu fugiat nulla pariatur. Excepteur sint occaecat cupiddeserunt mollit anim id est laborum. Lorem ipsum dolor sit amet, consectetur adipisicing elit, sed do eiusmod tempor incididunt ut labore et dolore magna aliqua.

Ut enim ad minim veniam

Quis nostrud exercitation ullamco laboris nisi ut aliquip ex ea commodo consequat. Duis aute irure non proident, sunt in culpa qui officia deserunt mollit anim id est laborum. fugiat nulla pariatur. Excepteur sint occaecat cupidatat non proident, sunt in culpa qui officia deserunt mollit anim id est laborum. Lorem ipsum dolor sit amet, consectetur adipisicing elit, sed do eiusmod tempor incididunt ut labore et dolore magna aliqua.

← Acid strength increases   7   Base strength increases →

The "strength" of an acid or base increases with distance from pH=7

pH of some common substances:

Acid	Neutral	Base
stomach acid—2	pure water—7	blood—7.5
cola drinks—3		sea water—8
tomatoes—4		detergent—10
coffee—5		household cleaners—11
milk—6.5		oven cleaners—14

Calculating pH:

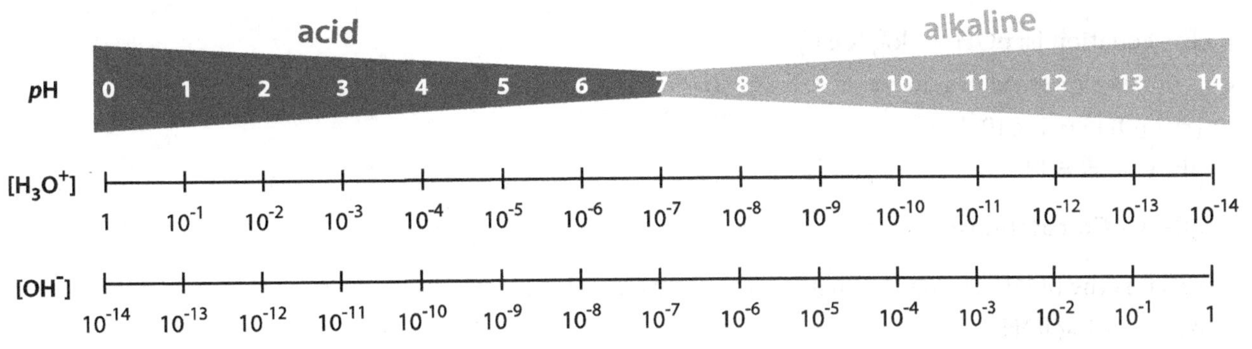

- The equation is: $pH = -\log [H_3O^+]$
- $[H^+]$ is expressed in powers of 10 from 10^{-14} to 10^0
- If $[H^+] = 1 \times 10^{-7}$, the negative log of $[H^+] = 7$. The pH equals 7, indicating a neutral solution.
- The calculation of pH always gives a number between 0 and 14.

Example pH Calculations:

Logarithm tables are needed to understand pH calculations. Scientific calculators have the log and anitlog functions.

1. What is the pH of a solution with a $[H^+]$ of 1.00×10^{-4} M?

   $pH = -\log [H^+]$

   $pH = -(\log 1.00 + \log 10^{-4})$

   $pH = -\log (1 \times 10^{-4})$

   $pH = -(0 + (-4))$

   $pH = -(-4)$

   $pH = 4$

2. 0.01 moles of HCl is added to water to make 1 dm3 of solution. Assuming the HCl is completely ionized, what is the pH of the solution?

   $[H^+] = 1 \times 10^{-2}$ M

   $pH = -\log [H_3O^+]$

   $pH = -\log (1 \times 10^{-2})$

   $pH = -(\log 1.00 + \log 10^{-2})$

   $pH = -(0 + (-2))$

   $pH = -(-2)$

   $pH = 2$

3. Calculate the $[H^+]$ of a solution with a pH of 3.70 using calculator:

   $pH = -\log [H^+]$

   $-pH = \log [H^+]$ . . . .

   $-3.70 = \log [H^+]$

   antilog $-3.70 = [H^+]$

   $[H^+] = 2 \times 10^{-4}$ M

pOH is a measure of the [OH -] in solution.

Calculating pOH:

- The equation is: pOH = –log $[OH^-]$
- $[OH^-]$ is expressed in powers of 10 from 10^{-14} to 10^0
- $[H^+]$ $[OH^-]$ = 1×10^{-14}
- pH + pOH = 14

Example pH Calculations:

1. What is the pOH of a solution with $[OH^-] = 3.98 \times 10^{-5}$ M?
   pOH = – log $[OH^-]$
   pOH = – log (3.98×10^{-5})
   pOH = – (log 3.98 + log 10^{-5})
   pOH = – (.5999 + (–5))
   pOH = – (–4.4)
   pOH = 4.40

2. Find the pH of a solution that contains 0.0035 moles of H^+
   pOH = – log $[OH^-]$
   pOH = – log (3.98×10^{-5})
   pOH = 4.40

Some terms

- pH meter:
  - An electronic device that measures pH directly.
  - pH meters are used in most professional lab settings today.
- Indicators:
  - Weak organic acids and bases whose colors differ from the colors of their conjugate acids or bases.
  - The color is best viewed from above against a white background.
  - Universal indicator solution has a wide range of color changes.
  - Hydrion paper:
    - A paper that goes through changes similar to the universal indicator solution.
  - Litmus paper:
    - Red litmus paper turns blue in a base.
    - Blue litmus paper turns red in an acid.
    - There is a litmus liquid with similar color response.

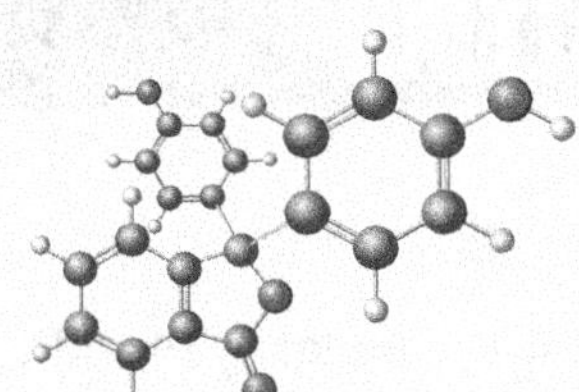

Phenolphthalein indicator

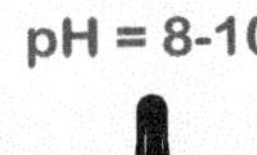

CHEMISTRY UNIT 9 PRACTICE PROBLEMS

MOLARITY

1. Sea water contains roughly 28.0 g of NaCl per 1.00 liter. What is the molarity of sodium chloride in sea water?
2. What is the molarity of 245.0 g of H_2SO_4 dissolved in 1.000 L of solution?
3. What is the molarity of 5.30 g of Na_2CO_3 dissolved in 400.0 mL solution?
4. How many moles of Na_2CO_3 are in 10.0 mL of a 2.0 M solution?
5. What mass (in grams) of H_2SO_4 would be needed to make 750.0 mL of 2.00 M solution?
6. What volume (in mL) of 18.0 M H_2SO_4 is needed to contain 2.45 g H_2SO_4?
7. What is the molarity of a solution made by dissolving 20.0 g of H_3PO_4 in 50.0 mL of solution?

DILLUTIONS

1. A stock solution of 1.00 M NaCl is available. How many milliliters are needed to make 100.0 mL of 0.750 M
2. What volume of 0.250 M KCl is needed to make 100.0 mL of 0.100 M solution?
3. Concentrated H_2SO_4 is 18.0 M. What volume is needed to make 2.00 L of 1.00 M solution?
4. Concentrated HCl is 12.0 M. What volume is needed to make 2.00 L of 1.00 M solution?

5. A 0.500 M solution is to be diluted to 500.0 mL of a 0.150 M solution. How many mL of the 0.500 M solution are required?

SOLUTIONS

1. Give an example of a homogeneous solution. Indicate the solute and solvent. Does this solution contain ions? If yes, indicate what the ions are. If no, how does the solute dissolve?

2. Illustrate how each of the following compounds dissolves in water.
   a. $BaCl_2$
   b. Ethanol, C_2H_5OH
   c. $Mg(ClO_4)$
   d. Perchloric acid, $HClO_4$ (a strong acid)

3. Explain the differences between saturated and super saturated solutions.

CONCENTRATION CALCULATIONS

1. Calculate the percent by mass of solute if 14.15 grams of NaI is mixed with 75.55 grams of water.

2. A solution contains 15.0 grams of NH_4Cl in water and 8.50% NH_4Cl. What is the mass of water present?

3. What is the molarity of a solution of 345 grams of Epsom salts ($MgSO_4 \bullet 7H_2O$) in 7.50 L of solution?

4. What is the volume in liters of a 0.440 *M* solution if it was made by dilution of 250 mL of a 1.25 *M* solution?

5. Calculate the grams of alcohol, C_2H_5OH, in 440 grams of a 23.0% solution.

6. 75.0 mL of water is added to 12.9 mL of 0.250 M $KMnO_4$. What is the concentration of the diluted solution?

7. If 2.54 g of sodium benzoate is dissolved in 75.00 g H_2O. What is the percentage concentration of sodium benzoate?

8. If 45.5 g of $BaCl_2$ is dissolved in water to produce 2.74 L of solution, what is the molarity of the solution?

9. Nitric acid is commercially available at a concentration of 15.9 M. What volume of this solution must be diluted to a final volume of 1.00 L to prepare a 4.00 M solution?

SOLUTION STOICHIOMETRY

1. Consider the reaction:
   a. $3\ Ca(ClO_3)_2(aq) + 2\ Na_3PO_4(aq) \rightarrow Ca_3(PO_4)_2(aq) + 6\ NaClO_3(aq)$
   b. What volume of a 2.22 *M* solution of Na_3PO_4 is needed to react with 580 mL of a 3.75 *M* solution of $Ca(ClO_3)_2$?

2. The citric acid in a lemon juice sample was neutralized by titration with NaOH solution. If 5.00mL of lemon juice required 47.8 mL of 0.121 M NaOH for neutralization, what was the molarity of the citric acid in the lemon juice?

   The reaction is ______ NaOH + ______ $H_3C_6H_5O_7$ → ______ H_2O + ______ $Na_3C_6H_5O_7$

IDENTIFYING ACIDS AND BASES

1. Identify each of the following as an acid or base when dissolved in water.

a. HClO	name
b. $Mn(OH)_2$	name
c. H_2Se	name
d. H_2SO_4	name
e. CuOH	name

2. Balance each of the following neutralization reactions.
   a. ______ HNO_3 + ______ $Ca(OH)_2$ → ______ $Ca(NO_3)_2$ + ______ H_2O
   b. ______ $HC_2H_3O_2$ + ______ NaOH → ______ $NaC_2H_3O_2$ + ______ H_2O
   c. ______ H_2SO_4 + ______ $Mg(OH)_2$ → ______ $MgSO_4$ + ______ H_2O

3. Identify the acid, base, conjugate acid, conjugate base for each of the following reactions.

a. $HClO_3$	+	H_2O	→	ClO_3^-	+	H_3O^+
b. HSO_4^-	+	NH_3	→	NH_4^+	+	SO_4^{2-}
c. HCO_3^-	+	H_2O	→	H_2CO_3	+	OH^-

pH AND pOH

1. What is the $[H^+]$ in the following solutions?
   a. $[OH^-] = 1.0 \times 10^{-12}$
   b. $[OH^-] = 5.9 \times 10^{-4}$
   c. pH = 9.0
   d. pH = 7.85
   e. pOH = 4.18

2. What is the pH of the following solutions?
   a. pOH = 4.51
   b. $[OH^-] = 1.0 \times 10^{-7}$
   c. $[H^+] = 3.6 \times 10^{-4}$

3. Which of the following is a strong acid?
   a. H_2CO_3
   b. $HC_2H_3O_3$
   c. HF
   d. HCl
   e. HClO

4. Which of the following is a characteristic property of a traditional acid?
   a. Turns litmus indicator blue
   b. Feels slippery
   c. Tastes bitter
   d. Sour taste
   e. Contain OH^- ions

5. Which one is a common indicator used in acid base titrations?
   a. sodium chloride
   b. propylene
   c. phenolphthalein
   d. bronsted-lowry
   e. arrhenius

6. According to Arrhenius theory, what is an acid?
   a. a substance which contains a high concentration of hydrogen ions in solutions with water
   b. a substance which will lower the hydrogen ion concentration when placed in water a substance that has an h in its formula
   c. an electron pair donor
   d. an electron pair acceptor

7. A Brønsted-Lowry acid is defined as a(n) _____.
   a. proton acceptor
   b. electron donor
   c. proton donor
   d. electron pair acceptor
   e. electron pair donor

8. What is a Brønsted-Lowry base?
   a. an electron pair donor
   b. a hydroxide ion donor
   c. a metal ion
   d. a proton donor
   e. a proton acceptor

9. What kind of substance is water in the following reaction?

   $H_2O(s) + HCN(aq) \rightarrow H_3O^+(aq) + CN^-(aq)$

   a. A Brønsted-Lowry base
   b. A Brønsted-Lowry acid

UNIT 1: PRACTICE PROBLEMS

DEFINING MATTER / STATES OF MATTER / PHYSICAL VS. CHEMICAL

Give an Example for Each of the Following:

1. Matter
   Matter is anything that has mass and occupies space. Break down matter and it will be comprised of one or more of the chemical elements. For examples use any material.

2. Macroscopic Matter
   Macroscopic matter is matter that can be seen with the human eye. List out any examples.

3. Microscopic Matter
   Microscopic matter is matter too small to be seen by the human eye, but can be seen with an optical microscope. List out any examples.

4. Particulate Matter
   Particulate matter is matter is the smallest piece of matter. When you break down matter to its smallest piece you have particulate matter. This includes an atom, a molecule, or a particle. List out any examples.

List the States of Matter, Describe, and Give an Example:

5. **Solid**

   Describe:
   Solids are rigid, relatively incompressible (into a different state of matter), have fixed volume and shape.

   Example:
   List any examples of solids

6. **Liquid**

Describe:

Liquids are wet, relatively incompressible (into a new state of matter), have a fixed volume but no fixed shape.

Example:

List any examples of liquids.

7. **Gas**

Describe:

Gases are easily compressible, have no fixed volume and no fixed shape.

Example:

List any examples of gases.

Give Two Examples for Each:

8. A physical property
color, density, freezing point, viscosity, smell... These are descriptions of just the material itself.

9. A chemical property
flammability, ability to oxidize, ability to react with water... These are descriptions of the ability or inability of material to react with other material.

10. A physical change
boiling, melting, freezing, dissolving, filtering... These are changes material undergoes but still remains to being the original material.

11. A chemical change
photosynthesis, cooking, oxidizing, ripening, fermenting, burning, exploding... These are changes material undergoes and becomes new material.

12. List observations that would show the differences between a physical change and a chemical change.
evidence of some chemical changes: gas released (not including a phase change like boiling or if the gas is already present like in soda), temperature change (without a heat source being added or removed), color change...

STATES OF MATTER / PHYSICAL VS. CHEMICAL

Fill in the Blanks.

1. **Physical** properties can be observed without chemically changing matter.
2. **Chemical** properties describe how a substance interacts with other substances.
3. **Solids** have definite shapes and definite volumes.
4. **Liquids** have indefinite shapes and definite volumes.
5. **Gases** have indefinite shapes and indefinite volumes.
6. **Freezing** point is the temperature at which a liquid turns to a solid. It is also equal to the **melting** point at which a **solid** turns to a **liquid**.
7. **Condensation** point is the temperature at which a gas turns to a liquid.
8. **Sublimation** is the occasional occurrence when a solid turns directly into a gas without turning into a liquid first.

Classify the following as Chemical Change (CC), Chemical Property (CP), Physical Change (PC), or Physical Property (PP).

Blue color	**PP**
Flammability	**CP**
Dissolves in water	**PP**
Scratches glass	**PP**
Rusting	**CC**
Reacts with oxygen to form a new compound	**CC**
Melting point	**PP**
Luster (shine)	**PP**

Density	**PP**
Heaviness	**PP**
Boils at 100 degrees	**PP**
Sour taste	**PP**
Exploding fireworks	**CC**
Reacts with H_2O to form gas	**CC**
Hardness	**PP**
Odor	**PP**

Heat conductivity	**PP**
Silver tarnishing	**CP**
Sublimation	**PC**
Magnetizing steel	**PC**
Length of metal object	**PP**
Melting	**PC**
Exploding dynamite	**CC**

Combustible	**CP**
Water freezing	**PC**
Wood burning	**CC**
Acid resistance	**PP**
Brittleness	**PP**
Milk souring	**CC**
Baking bread	**CC**

Identify the following as being true or false

TRUE 1. A change in size or shape is a physical change.
TRUE 2. A chemical change means a new substance with new properties was formed.
FALSE 3. An example of a chemical change is when water freezes.
TRUE 4. When platinum is heated, then cooled to its original state, we say this is a physical change.
FALSE 5. When water turns sour, this is a physical change because a change in odor does not indicate a chemical change.
TRUE 6. When citric acid and baking soda mix, carbon dioxide is produced and the temperature decreases. This must be a chemical change.

Identify each of the following as a physical or chemical change.

__CC__ 1. You leave your bicycle out in the rain and it rusts.
__PC__ 2. A sugar cube dissolves.
__CC__ 3. Scientists break up water into oxygen and hydrogen gas.
__CC__ 4. Coal is burned for a barbecue.
__PC__ 5. A shrub or tree is trimmed because it has grown too tall.

CLASSIFYING MATTER

1. According to the demonstration presented in class, write a chemical reaction and indicate what happens to the mass before and after the reaction.

   answer given in class.

2. Describe and explain the particulate level of an element compared to a compound.

   When a single element like copper is broken down to its smallest piece of matter you will have an atom of copper. When a compound like water is broken down to its smallest piece of matter you will have a molecule of water comprised of one atom of oxygen bonded to two atoms of hydrogen. This bond is created by the sharing of electrons between the atoms so we call this piece of matter a molecule. When a compound like table salt, also called sodium chloride is broken down to its smallest piece of matter you will have a particle (or formula unit) of sodium chloride comprised of one sodium atom bonded to one chlorine atom. This bond is created by the transfer of electrons between the atoms so we call this piece of matter a particle (or formula unit).

3. Describe and give an example for each of the following:
   a. Heterogeneous
      a. Describe:

         Heterogeneous are physical combinations of materials that are not uniformly mixed.

      b. Example:

         oil and water, smog, tree, sand

   b. Homogeneous / Solution
      a. Describe:

         Homogeneous Solutions are physical combinations of materials that mix uniformly.

      b. Example:

         clean sea water or drinking water, brass (metal alloys), clean air

c. Homogeneous / Pure Substance / Compound

a. Describe:

Homogeneous pure substance compounds are material composed of more than one element that have chemically reacted via the sharing of electrons or the transfer of electrons. To identify these you will see a chemical formula written.

b. Example:

carbon dioxide, water, sucrose, sodium chloride, sodium bicarbonate, ethanol

d. Homogeneous / Pure Substance / Element

a. Describe:

Homogeneous pure substance elements are only comprised of one of the chemical elements found on the periodic table. (Note the diatomics are molecules and therefore would be classified as a compound.)

b. Example:

copper, sodium, tin, sulfur, neon

4. Fill in the blank
   a. Sb **antimony**
   b. Cu **copper**
   c. Au **gold**
   d. Fe **iron**
   e. Pb **lead**
   f. Hg **mercury**
   g. Na **sodium**
   h. W **tungsten**
   i. K **potassium**
   j. Ag **silver**
   k. Sn **tin**

5. List the diatomic molecules

   I_2, H_2, N_2, Br_2, O_2, Cl_2, F_2

CLASSIFYING MATTER

Classify each of the materials below:

1. Heterogeneous
2. Homogeneous / Solution
3. Homogeneous / Pure Substance / Compound
4. Homogeneous / Pure Substance / Element

Material	1, 2, 3, or 4
Powder laundry detergent (contains white and blue crystals)	**1**
sugar + pure water ($C_{12}H_{22}O_{11} + H_2O$)	**2**
Iron Filings (Fe)	**4**
limestone ($CaCO_2$)	**3**
orange juice (water and pulp)	**1**
Pacific Ocean (water and salt)	**2**
air	**2**
aluminum (Al)	**4**
magnesium (Mg)	**4**
acetylene (C_2H_2)	**3**
tap water in glass	**2**
pure water (H_2O)	**3**
soil	**1**
chromium (Cr)	**4**
baking soda (sodium bicarbonate)	**3**
salt + pure water ($NaCl + H_2O$)	**2**
benzene (C_6H_6)	**3**
muddy water	**1**
brass (Cu mixed with Zn)	**2**
Pizza	**1**

Material	1, 2, 3, or 4
Spicy salad dressing	**1**
95% Rubbing alcohol (C_2H_8O + water)	**2**
Captain Morgan (80% Vodka and 20% water)	**2**
compost pile	**1**
gold necklace (gold and copper)	**2**
Smog (dirty air)	**1**
lye (NaOH)	**3**
Nail polish remover (C_2H_6O + water)	**2**
Plaster of Paris (calcium sulfate)	**3**
Opened soda pop	**1**
100% Rubbing alcohol (C_2H_6O)	**3**
Chocolate chip ice cream	**1**
Diamond (carbon)	**4**
Steel (alloy of iron + other elements)	**2**
Black ink (comination of colors)	**2**
Carbon dioxide gas (CO_2)	**3**
Floor tile	**1**
Human tear	**2**
Pre-1982 penny (95% Cu and 5% Zn)	**2**
Blueberry bagel	**1**

ENERGY

Part 1. Definitions of energy.

Write down the definition for each of the following terms:

Energy:
Energy is the ability to do work. Matter contains stored energy that is released in its movement.

Kinetic energy:
Kinetic energy is the energy associated with motion and is the energy matter releases in its movement.

Potential energy:
Potential energy is the energy that is stored as a function of position and is the energy matter stores.

Part 2. The two basic types of energy.

**Determine the best match between basic types of energy and the description provided.
Put the correct letter in the blank.**

(a) Kinetic Energy
(b) Potential Energy

__B__ 1. A skier at the top of the mountain
__B__ 2. Gasoline in a storage tank
__A__ 3. A race-car traveling at its maximum speed
__A__ 4. Water flowing from a waterfall before it hits the pond below
__B__ 5. A spring in a pinball machine before it is released
__A__ 6. Burning a match
__A__ 7. A running refrigerator motor

Part 3. Forms of Energy.

Determine the type of energy for each form (Kinetic or Potential).

Form	Definition	Type
Mechanical (motion) energy	An object's movement creates energy	**Kinetic**
Thermal (heat) energy	The vibration and movement of molecules	**Kinetic**
Electrical energy	Movement of electrons	**Kinetic**
Chemical energy	Stored in bonds of atoms and molecules	**Potential**
Nuclear energy	Stored in the nucleus of an atom; released when nucleus splits or combines	**Potential/ Kinetic**

Endothermic Reactions vs. Exothermic Reactions

Classify each of the following changes as either exothermic or endothermic. Explain your reasoning for each.

Process	Exo	Endo	Explanation
An ice cube melts after being left out on the table.		**X**	
Cooking an egg in a frying pan.		**X**	
Burning a match.	**X**		
The human body uses the energy provided from food digestion.		**X**	
Morning dew forming on grass and plants.	**X**		
Dynamite explodes in the destruction of a building.	**X**		
Making ice cubes.	**X**		
A puddle of water evaporates.		**X**	
Plants making sugar through photosynthesis.		**X**	
Converting frost to water vapor.		**X**	
Your own examples			

REVIEW UNIT 1:

__c__ 1. Which of the following correctly matches the classification as microscopic?
a. teeth
b. finger nail
c. a red blood cell
d. cat hair
e. a spoonful of sugar

__d__ 2. Which of the following does *not* describe the liquid state?
a. wet and takes the shape of its container
b. can measure its volume
c. can be frozen to a solid
d. usually has a lower density than a gas
e. all describe the liquid state

__e__ 3. Which of the following describes a chemical change?
a. dissolving
b. evaporating
c. melting
d. freezing
e. burning

__e__ 4. Which of the following is a solid at room temperature?

a. mercury
b. hydrogen
c. oxygen
d. neon
e. silicon

__a__ 5. Which of the following is a chemical change?

a. rusting
b. melting
c. freezing
d. boiling
e. dissolving

__e__ 6. Which of the following changes is/are classified as physical?

i. carving a block of ice into a sculpture
ii. iron rusting
iii. sugar dissolving in water
iv. A candle burning
v. souring of cream

a. i only
b. v only
c. i and v
d. i, iii, and v
e. i and iii

__b__ 7. Breaking water up by separating it into hydrogen and oxygen is an example of a _____.

a. physical change
b. chemical change
c. liquid changed to solid
d. solid changed to liquid

__a__ 8. Which of the following is a chemical property?

a. combustibility
b. boiling point
c. density
d. odor
e. taste

__b__ 9. Which of the following is a physical property of magnesium?

a. reacts with bromine
b. the color is silver
c. reacts with chlorine
d. reacts with oxygen
e. burns easily

__c__ 10. Which of the following is classified as a pure substance?

a. wine
b. fog
c. acetone
d. sand
e. milk

__a__ 11. Classify the material concrete.

a. heterogeneous
b. homogeneous / solution
c. heterogeneous / pure substance / compound
d. homogeneous / pure substance / compound
e. homogeneous / pure substance / element

__d__ 12. Classify the material 100% hydrogen peroxide.

a. heterogenous / solution
b. homogeneous / solution
c. heterogeneous / compound
d. homogeneous / compound
e. heterogeneous / element

__d__ 13. Classify the material vodka (alcoholic beverage).

a. heterogeneous
b. homogeneous / pure substance / compound
c. heterogeneous / pure substance / compound
d. homogeneous / solution
e. heterogeneous / solution

__a__ 14. Which of the following is classified as a pure substance?

a. sodium
b. brass
c. milk
d. iced tea
e. soda

__e__ 15. Which of the following is a heterogeneous mixture?

a. nitrogen gas
b. filtered air
c. silver
d. alcohol
e. sand

__a__ 16. Which of the following correctly describes a homogeneous sample?

a. uniform appearance and composition throughout
b. visibly different parts or phases
c. a mixture of water and oil
d. contains no elements
e. contains no compounds

__d__ 17. Which of the following substances is homogeneous / pure substance / compound?

a. a tree branch
b. concrete
c. smog (dirty air)
d. sugar
e. salad dressing

__b__ 18. Which of the following substances is heterogeneous?

a. 3% hydrogen peroxide
b. the lint from your clothes dryer
c. a copper wire
d. baking soda
e. iron

__c__ 19. What type of substance cannot be decomposed or separated into other stable pure substances?

a. a gas like propane
b. a solution like salt water
c. an element like nickel
d. a compound like CH_4
e. homogeneous matter like 90% rubbing alcohol

__d__ 20. Which of the following is *not* a diatomic molecule?

a. hydrogen
b. oxygen
c. chlorine
d. sulfur
e. nitrogen

__b__ 21. Which of the following is an exothermic changes.

a. a cold pack
b. a hot pack

__b__ 22. When paper is burned, the mass of the remaining ash is less than the mass of the original paper. Does the behavior happen with all material when it burns?

a. yes
b. no

HISTORY

1. What are the four elements in Aristotle's theory of matter?
   Earth, fire, air, water

2. What is the philosopher's stone and how does it relate to early chemistry?
   The philosopher's stone was a myth believed to transmute metal and bring eternal life.

3. Define alchemy?
   A pseudoscientific forerunner of chemistry in medieval times.

4. Where did many of our earliest understandings of chemistry evolve?
   Greek philosophers

5. What is phlogiston? Does it exist?
   Phlogiston was a substance proposed by Becher as a reason for substance changes when burned. Phlogiston does not exist.

6. Who disproved the phlogiston theory and gave oxygen its name?
   Lavoisier

7. According to the phlogiston theory, what is the carrier of phlogiston?
   Air

MATTER

1. Define matter.
   Matter is anything that has mass and occupies space.

2. Discuss how a sample of sugar could be classified as:
   a. Macroscopic material
      A teaspoon of sugar

   b. Microscopic material
      A crystal of sugar

   c. Particulate material
      A molecular of sugar ($C_{12}H_{22}O_{11}$)

3. Who proposed the four elements of matter are earth, water, fire, and air?
   Aristotle

4. Classify the following as macroscopic, microscopic, or particulate: a carbon dioxide molecule.
   Particulate

5. Classify the following as macroscopic, microscopic, or particulate: a red blood cell.
   Microscopic

6. The word pour is commonly used in reference to liquids, but not to solids or gases. Can a solid or gas be poured? Why or why not? If either answer is yes, can you give an example?
   Gases can be poured (there are gases that are heavier than air).

7. Is it possible to melt a piece of ice without seeing a change in temperature? Explain.
   Yes, in fact, substances melt (go from solid to liquid) without a temperature change. At 0°C, ice will turn to liquid water with an increase in heat (energy) and no temperature change.

8. Which state of matter (solid, liquid, or gas) has no atoms moving?
   All states of matter have atoms moving.

PHYSICAL AND CHEMICAL PROPERTIES AND CHANGES

1. Classify the following as chemical or physical properties.
    a. the mass of a chunk of nickel is 19 grams **Physical**
    b. when sulfur is exposed to water it burns **Chemical**
    c. the color of copper is a rust-like color **Physical**
    d. sugar can dissolve in water **Physical**
2. Classify the following changes as chemical or physical.
    a. rusting car **Chemical**
    b. boiling water **Physical**
    c. digesting rice in the stomach **Chemical**
    d. dropping Alka Seltzer™ in water **Chemical**
3. Which of the following is an example of a chemical change?
    a. water boiling
    b. ice melting
    c. natural gas burning
    d. iodine vaporizing
    e. dry ice turning from the solid phase to the gas phase
4. List some chemical properties of water?
    It can be separated into hydrogen and oxygen by electrolysis, it forms bubbles when calcium is added, etc.

CLASSIFYING MATTER

1. The smallest particle of an element that retains the chemical properties of the element is a(n)?
    Atom
2. To what category would a sample containing more than one phase belong?
    a. compound
    b. element
    c. diatomic mixture
    d. homogeneous mixture
    e. heterogeneous mixture
3. Which of the following is a homogeneous solution?
    a. vinegar (5% acetic acid)
    b. water
    c. baking soda
    d. sewage
    e. diamond
4. Classify boron, carbon dioxide, methane, and lead as element or compound.
    Compound: carbon dioxide, methane
    Element: boron, lead
5. What is the symbol for Antimony?
    Sb

6. List the diatomic molecules.

   Hydrogen, iodine, bromine, oxygen, nitrogen, chlorine, iodine, fluorine (HI BrONClIF)

7. Which of the following is a pure substance?
   a. vinegar (5% acetic acid)
   b. concrete
   c. baking soda
   d. sewage
   e. brass

8. How would you classify steel?
   a. heterogeneous mixture
   b. homogeneous, solution
   c. homogeneous, compound
   d. homogeneous, element

9. What is the chemical symbol for mercury?

   Hg

10. Which of the following is a physical combination of two or more pure substances?
    a. salt
    b. air
    c. sand
    d. water
    e. natural gas

11. Classify a mixture of 80% nitrogen (N_2) and 20% oxygen (O_2) which can be separated into these compounds by cooling (a physical change) is commonly called?

    Air

12. What does the following chemical formula indicate? Al_2O_3
    a. 3 atoms aluminum, 2 atoms oxygen
    b. 1 atom aluminum, 1 atom oxygen
    c. 1 atom aluminum oxygen
    d. 6 atoms aluminum oxygen
    e. 2 atoms aluminum, 3 atoms oxygen

13. Which of the following is a chemical compound?
    a. Co
    b. CO
    c. Pb
    d. Sn
    e. Ar

14. Which of the following is a chemical compound?
    a. alcohol
    b. aluminum
    c. antimony
    d. argon
    e. astatine

15. How would you classify soda pop?
    a. heterogeneous mixture
    b. homogeneous, solution (unopened soda pop)
    c. homogeneous, compound
    d. homogeneous, element

16. How would you classify propane?
    a. heterogeneous
    b. homogeneous, solution
    c. homogeneous, pure substance

17. Which of the following is a diatomic molecule?
    a. Ar_2
    b. Br_2
    c. Cr_2
    d. Dr_2
    e. Er_2

18. Write the chemical name or symbol for each of the following:

Krypton **Kr**	Al **Aluminum**	Chromium **Cr**
Carbon **C**	Cl **Chlorine**	Tellurium **Te**
Mercury **Hg**	P **Phosphorus**	Protactinium **Pa**

19. Indicate which category or categories each of the following substances belongs.

Substance	Heterogeneous	Homogeneous	Solution	Pure Substance	Element	Compound
Pure NaCl		X		X		X
Al foil		X		X	X	
Filtered air		X	X			
Concrete	X					
Sand	X					
Plastic (polystyrene)		X		X		X
Astatine		X		X	X	
Salt water		X	X			

CHEMICAL REACTIONS & ENERGY

1. Write the chemical equation for the following described reactions. Identify the reactants and products (It is alright if your equation is not balanced, we will address this topic in a future unit).
    a. Propane (C_3H_8) burns (in the presence of oxygen) to produce carbon dioxide (CO_2) and water.
    $\mathbf{C_3H_8}(l) + \mathbf{5O_2}(g) \rightarrow \mathbf{3CO_2}(g) + \mathbf{4H_2O}(g)$

    b. Energy is required to separate NaCl into its elements.
    $\mathbf{2NaCl}(s) + \mathbf{energy} \rightarrow \mathbf{2Na}(s) + \mathbf{Cl_2}(g)$

2. Classify the following as kinetic or potential energy.
   a. a running halfback **Kinetic energy**
   b. a bowling ball at the top of the stairs **Potential energy**
   c. water flowing over the falls **Kinetic energy**
   d. a moving train **Kinetic energy**
   e. a strap holding down a bike on a trailer **Potential energy**

3. Classify the following as an endothermic or exothermic reaction.
   a. ice melting **Endothermic**
   b. wood burning **Exothermic**
   c. an ice pack (for sports injuries) **Endothermic**
   d. a heating pad **Exothermic**

UNIT 2: PRACTICE PROBLEMS

MEASUREMENT / SCIENTIFIC NOTATION / METRIC CONVERSIONS

Based upon the following pictures:

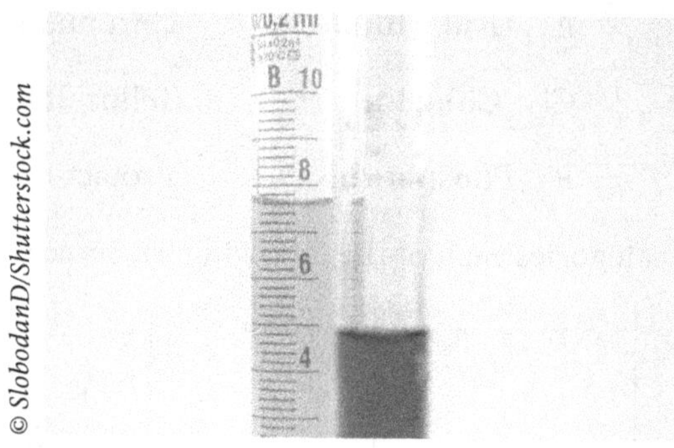

© SlobodanD/Shutterstock.com

1. How would you record the volume of this liquid? **6.80 mL**

© Basileus/Shutterstock.com

2. How would you record the length of this die? **The recorded length is based upon the precision of the instrument.**

3. Write the following numbers in scientific notation:

a. 321,012,000	a.	3.21012×10^8
b. 0.0000330000	b.	3.30000×10^{-5}
c. 0.000098500	c.	9.8500×10^{-5}
d. 4.67	d.	4.67×10^0

4. Write the following numbers in ordinary decimal notation:

a. 4.32×10^6	a.	**4,320,000**
b. 4.32×10^{-2}	b.	**0.0432**
c. 9.800×10^{-7}	c.	**0.0000009800**
d. 4.006000×10^9	d.	**4,006,000,000**

5. Convert the following units:

a. 5.68 kg to mg	a.	5.68×10^6 mg
b. 0.000987 dm to mm	b.	**0.0987 mm**
c. 8.76×10^4 mL to L	c.	8.67×10^1 L
d. 5.4×10^{-4} μs to ms	d.	5.4×10^{-7} ms
e. 9.56×10^9 dL to kL	e.	9.56×10^5 kL
f. 78.998 mg to ng	f.	7.8998×10^7 ng
g. 0.0087 ns to cs	g.	8.7×10^{-10} cs
h. 8.99×10^{-4} cm³ to μL	h.	8.99×10^{-1} microliter

Write the following number into scientific notation:

1. 100,000,000,000 = $\mathbf{1 \times 10^{11}}$	2. 0.0034 = $\mathbf{3.4 \times 10^{-3}}$
3. 699,000,000 = $\mathbf{6.99 \times 10^8}$	4. 0.0000597 = $\mathbf{5.97 \times 10^{-5}}$
5. 3,003 = $\mathbf{3.003 \times 10^3}$	6. 0.00908070 = $\mathbf{9.08070 \times 10^{-3}}$

Write the following number into ordinary decimal notation:

7. 1.0×10^8 = **100,000,000**	8. 2.3×10^3 = **2300**
9. 77×10^5 = **977,000**	10. 8.76×10^{-4} = **0.000876**
11. 9.3×10^{-3} = **0.0093**	

Convert the following:

12. 10 km = **1,000,000 cm** = $\mathbf{1 \times 10^6}$ **cm**	13. 1×10^6 mL = $\mathbf{1 \times 10^3}$ **L = 1,000 L**
14. 1×10^9 kg = $\mathbf{1 \times 10^{18}}$ **μg**	15. 100 m = **0.1 km** = $\mathbf{1 \times 10^{-1}}$ **km**
16. 1×10^4 ng = $\mathbf{1 \times 10^{-2}}$ **mg = 0.01 mg**	17. 6.5 μm = $\mathbf{6.5 \times 10^{-5}}$ **dm**
18. 4.56 dg = **456 mg** = $\mathbf{4.56 \times 10^2}$ **mg**	19. 0.689 ks = **68900 cs** = $\mathbf{6.89 \times 10^4}$ **cs**
20. 0.00223 μL = **0.00000223 mL** = $\mathbf{2.23 \times 10^{-6}}$	21. 5,555,000 nm = **0.05555 dm**
22. 3.58 ms = $\mathbf{3.58 \times 10^6}$ **ns**	

Determine the number of significant digits in each of the following:

23. 5.40 = **3**	24. 1.2×10^3 = **2**
25. 210 = **2**	26. 0.00120 = **3**
27. 801.5 = **4**	28. 0.0102 = **3**
29. 1,000 = **1**	30. 9.010×10^{-6} = **4**
31. 101.0100 = **7**	32. 2,370.0 = **5**

Add:

33. 16.5 + 8 + 4.37 = **29**	34. 13.25 + 10.00 + 9.6 = **32.9**
35. 2.36 + 3.38 + 0.355 + 1.06 = **7.16**	36. 0.0853 + 0.0547 + 0.0370 + 0.00387 = **0.1809**
37. 25.37 + 6.850 + 15.07 + 8.056 = **55.35**	38. 2390 + 123.9 = **2510**

Subtract:

39. 23.27 – 12.058 = **11.21**	40. 13.57 – 6.3 = **7.3**
41. 27.68 – 14.369 = **13.31**	42. 350.0 – 200 = **200**

Multiply:

43. 2.6×3.78 = **9.8**	44. 3.08×5.2 = **16**
45. 6.54×0.37 = **2.4**	46. 0.0036×0.02 = **0.00007** = **7×10^{-5}**

Divide:

47. 45.00/8.9 = **5.1**	48. 0.09/0.31 = **0.3**
49. 8.6700/0.002 = **4,000**	50. 34.005/0.0098 = **3,500**

MEASUREMENT / SCIENTIFIC NOTATION / METRIC CONVERSIONS

1. Describe how each is measured and give a SI unit for each.
   a. mass
      measured with a balance, SI unit is the kilogram, Kg (base unit is gram, g)

   b. length
      measured with a meter stick, SI unit is the meter, m

   c. time
      measured with a stopwatch, SI unit is the second, s

   d. temperature
      measured with a thermometer, SI unit is the Kelvin, K (measurement taken in Celsius)

   e. volume
      measured with a graduated cylinder, SI unit is the liter, L

2. Convert the following units:
   a. 4.56 dg = **456 mg**
   b. 0.00223 μL = **2.23×10^{-6} mL**
   c. 0.689 ks = **68900 cs**
   d. 5,555,000 nm = **0.05555 dm**

3. Determine the number of significant figures in each of the following numbers:
   a. 54.0003 mL = **6**
   b. 0.0003200 cm^3 = **4**
   c. 34.003 mL = **5**
   d. 45,500 cm = **3**

4. Round the following numbers to 3 significant figures:
   a. 34,690 = **34,700**
   b. 0.4500000 = **0.450**
   c. 210,000 = **2.10×10^5**

5. Calculate the following and record your answer with the proper number of significant figures:
   a. 54.900 + 32.1 = **87.0**
   b. (65.00 × 0.21) / 0.03 = **500**
   c. 43,000 × 65.00 = **2,800,000**
   d. (34.56 – 34.01) × (23.4 + 11.1) = **19**

6. Convert the following and record your answers in scientific notation:
   a. 0.225 dm = **2.25×10^1 mm**
   b. 44,163 cs = **4.4163×10^{-1} ks**
   c. 0.00000000000991 kL = **9.91×10^{-3} μL**
   d. 2.5 mg = **2.5×10^6 ng**
   e. 20,190 μs = **2.019×10^1 ms**
   f. 7,000 nL = **7×10^{-9} kL**

7. How many significant figures are in each of the following numbers?
   a. 5.40 km = **3**
   b. 210 μL = **2**
   c. 801.5 g = **4**
   d. 1,000 ns = **1**
   e. 101.0100 mL = **7**
   f. 1.2×10^3 mm = **2**
   g. 0.00120 cg = **3**
   h. 0.0102 ds = **3**
   i. 9.010×10^{-6} ks = **4**
   j. 2,370.0 μg = **5**

8. Round off each of the following numbers to four significant figures:
   a. 15.9994 nL = **16.00 nL**
   b. 0.66549 s = **0.6655 s**
   c. 87,500 mg = **8.750×10^4 mg**
   d. 1.0080 kL = **1.008 kL**
   e. 48,859 μm = **48,860 μm**
   f. 0.027225 s = **0.02723 s**

9. Calculate the following, answer using proper significant figures and units:
   a. 43.00 dL + 19.1 dL = **62.1 dL**
   b. (13.980 cg – 11.0 cg) + (340 cg + 217 cg) = **560 cg**
   c. 2.7800×10^4 nm + 1.10×10^4 nm = **3.88×10^4 nm**
   d. 0.00450 mm × 3.0454 mm × 1.0000 mm = **0.0137 mm³**
   e. $\dfrac{0.0320m \times 9.100m \times 0.0045m}{4320m}$ = **3.0×10^{-7} m²**
   f. $\dfrac{(4230g + 547g)}{0.1340cm \times 0.050cm \times 3.10cm}$ = **230000 g/cm³**
   g. (8.900×10^8 km) × (3.20×10^{-5} km) = **28500 km²**

DENSITY

Answer the following questions. Remember to include units and significant figures in your answer.

1. Calculate the density of each of the following:
   a. 252 mL of a solution with a mass of 500 g
      1.98 g/mL

   b. A 6.75 g solid with a volume of 5.35 cm³
      1.26 g/cm³

   c. A substance with a mass of 7.55×10^4 kg and a volume of 9.50×10^3 L
      7.95 kg/L or 7.95 g/mL

2. Calculate the volume of each of the following:
   a. 26.5 g of a solution with a density of 7.48 g/mL
      3.54 mL

   b. A 3.400 kg solid with a density of 10.74 g/mL
      316.6 mL

3. Calculate the mass of each of the following:
   a. A solid with a volume of 1.68 L and a density of 9.2 g/mL
      15456 g = 15000 g = 15 kg

   b. An 80 mL aliquot of a solution with a density of 5.80 g/cm^3
      464 g = 500 g

   c. A solid cube with a density of 2.65 g/mL and dimensions of 0.50 m × 250 mm × 3.5 cm
      11594 g = 12000 g = 12 kg

SI BASE UNITS

1. What are some of the advantages to having a SI (metric) system of measurement?
   To use an example:
   In the English system instead of saying 18 inches we can say 1 ½ feet. Inches and feet are both length units in the same system.
   The SI system is not so complex. Instead of saying 120 centimeters, we can say 1.2 meters. Meters and centimeters are both units in the same system. Here it is easier to see their relationship. Also, the SI system is a universal system of measurement.

2. Write out a unit of measurement that is not used in the SI (metric) system.
   Inches, pounds, yards, etc

3. What is the SI base unit for length?
   Meter

4. What is mass measured with?
   Balance

5. Is volume a derived unit?
   Yes

6. What is the SI base unit for volume?
   Liter

7. Determine the SI base units for the following:

a. time	**second—s**
b. temperature	**Kelvin—K**
c. length	**meter—m**
d. mass	**mass—kg**
e. amount of substance	**mole—mol**

8. Describe each (how it is measured) and give a SI unit for each:

a. mass	**balance, gram (or kg)**
b. length	**ruler, centimeter**
c. time	**stopwatch, second**
d. temperature	**thermometer, Kelvin**
e. volume	**grad. cylinder, liter**

SCIENTIFIC NOTATION

1. Change the following into scientific notation:
   a. 123,000,000,000,000 = **1.23×10^{14}**
   b. 234,044,000 = **2.34044×10^{8}**
   c. 0.00000034 = **3.4×10^{-7}**
   d. 0.001002 = **1.002×10^{-3}**

2. Change the following into ordinary decimal notation:
   a. 5.432×10^{-5} = **0.00005432**
   b. 3.21×10^{1} = **32.1**
   c. 4.56×10^{12} = **4,560,000,000,000**
   d. 8.99×10^{-2} = **0.0899**

3. Change the following into ordinary decimal notation:
   a. 5.67×10^{6} = **5,670,000**
   b. 3.4003×10^{-3} = **0.0034003**
   c. 8.8900×10^{-7} = **0.00000088900**
   d. 7.71×10^{1} = **77.1**

4. Change the following into scientific notation:
   a. 0.0450000 = **4.50000×10^{-2}**
   b. 3.21 = **3.21×10^{0}**
   c. 45,800,000 = **4.58×10^{7}**
   d. 230.01 = **2.3001×10^{2}**
   e. 0.00056700 = **5.6700×10^{-4}**

5. Write 7.19×10^{-4} in decimal notation.
   0.000719

6. Write 8.34×10^{-3} in decimal notation.
   0.00834

METRIC CONVERSIONS

1. Match the following:

i. 1 milligram	**c**	a.	1×10^{-2} gram
ii. 1 centigram	**a**	b.	1×10^{3} gram
iii. 1 kilogram	**b**	c.	1×10^{-3} gram
iv. 1 decigram	**d**	d.	1×10^{-1} gram
v. 1 microgram	**e**	e.	1×10^{-6} gram
vi. 1 nanogram	**f**	f.	1×10^{-9} gram

2. Complete the following:
   a. 1 kilometer = **1000** meters
   b. 1 meter = **10** decimeters
   c. 1 meter = **100** centimeters
   d. 1 meter = **1000** millimeters
   e. 1 meter = $\mathbf{1 \times 10^6}$ micrometers
   f. 1 meter = $\mathbf{1 \times 10^9}$ nanometers

3. How many milligrams are in 1 gram?
   1000 mg

4. How many decimeters are in 1 meter?
   10 dm

5. What is larger 1000 dg or 10 kg?
   10 kg

6. What is smaller 1 mL or 10 L?
   1 mL

7. Determine the following conversions (using dimensional analysis):
   a. How many kilograms are there in 50.0 grams?

$$\mathbf{50.0 \mathit{grams} \times \frac{1\mathit{kg}}{1000\mathit{g}} = 0.0500\mathit{kg}}$$

   b. How many mL are there in 4.58 L?

$$\mathbf{4.58\mathit{L} \times \frac{1000\mathit{mL}}{1\mathit{L}} = 4580\mathit{mL}}$$

   c. How many micrometers are there in 0.0000563 millimeters?

$$\mathbf{0.0000563\mathit{mm} \times \frac{1\mathit{m}}{1000\mathit{mm}} \times \frac{1 \times 10^6 \mu\mathit{m}}{1\mathit{m}} = 0.0563\mu\mathit{m}}$$

d. How many deciseconds are there in 56,700 kiloseconds?

$$56{,}700ks \times \frac{1000s}{1ks} \times \frac{10ds}{1s} = 5.67 \times 10^{8} ds$$

8. Determine the following conversions (using dimensional analysis):
   a. 6.7 g = **6.7×10^{-3} kg**
   b. 545.4 mm = **54.54 cm**
   c. 0.0458 ms = **45.8 μs**
   d. 5,987 m = **59,870 dm**
   e. 0.0303 mm = **3.03×10^{-8} km**
   f. 0.678 kg = **67,800 cg**
   g. 3.49 ng = **3.49×10^{-3} μg**

9. Convert the following and record your answers in scientific notation:
   a. 0.225 dm to mm = **2.25×10^{1} mm**
   b. 44,163 cs to ks = **4.4163×10^{-1} ks**
   c. 0.00000000000991 kL to μL = **9.91×10^{-3} μL**
   d. 2.5 mg to ng = **2.5×10^{6} ng**
   e. 20,190 μs to ms = **2.019×10^{1} ms**
   f. 7,000 nL to kL = **7×10^{-9} kL**

10. The unaided human eye has a resolving power of 0.1 mm. What is the equivalent resolving power in micrometers?

    1×10^{2} μm

SIGNIFICANT FIGURES

1. A crucible is known to weigh 24.3162 g. Three students in the class determine the mass of the crucible by repeated massings on a simple balance. Using the following information, which student has done the most precise determination?

Student	Trial 1	Trial 2	Trial 3	Trial 4	Trial 5
A	24.8	24.9	24.8	24.9	24.8
B	24.8	24.0	24.2	24.1	24.3
C	24.5	24.1	24.5	24.1	24.3

Student A has done the most precise work (Student C is the most accurate).

2. Determine the number of significant figures in the following:
   a. 234.0012 = **7 sig figs**
   b. 0.004302 = **4 sig figs**
   c. 34.000 = **5 sig figs**
   d. 45.67900 = **7 sig figs**
   e. 300 = **3 sig figs**
   f. 3.200 = **4 sig figs**

3. How many significant figures are in 0.004050000?

   7 sig. figs.

4. When measuring the length of this line with the metric ruler provided, the first decimal place that is uncertain is:

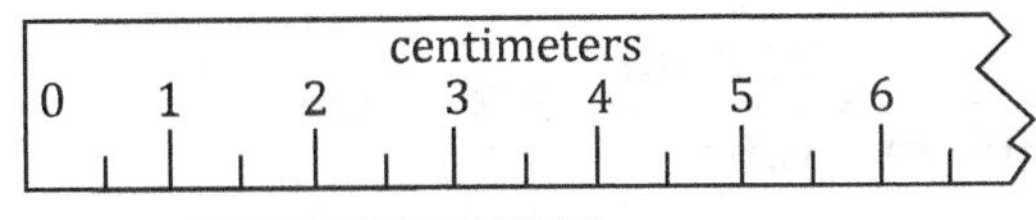

Tenths of a centimeter. (4.0 cm)

5. How many significant figures are in the measurement 102.400 meters?
6 sig. figs.

6. The mass of a watch glass was measured four times. The masses were 99.997 g, 100.008 g, 100.011 g, and 100.005 g. What is the average mass of the watch glass?
100.01 g

7. Solve to the correct significant figures: 1.23 m × 0.89 m = ?
1.1 m^2

8. Round the following measurement to three significant figures: 0.90985 cm^2
0.910 cm^2

9. Solve to the correct significant figures: 3.12 g + 0.8 g + 1.033 g = ?
5.0 g

10. When performing the calculation 34.530 g + 12.1 g + 1,222.34 g, how many significant figures will the final answer have?
5 sig. figs. (1268.97 rounded to 1269.0)

11. Perform each calculation and express all answers in the correct number of significant figures:
    a. $9.40\ cm \times 2.6\ cm$

    $$\mathbf{9.4\overset{3}{0}\ cm \times 2.\overset{2}{6}\ cm = 24.44\ cm = 24\ cm}$$

    b. $\dfrac{0.084\,g}{0.640\,mL}$

    $$\mathbf{\frac{0.0\overset{2}{8}4\,g}{0.6\underset{3}{4}0\,mL} = 0.13125\,g/mL = 0.13\,g/mL}$$

c. $\dfrac{21.50\,g}{4.06\,cm \times 1.8\,cm \times 0.905\,cm}$

$$\mathbf{\frac{\overset{4}{21.50}g}{\underset{3}{4.60}cm \times \underset{2}{1.8}cm \times \underset{3}{0.905}cm} = \frac{\overset{4}{21.50}g}{\underset{2}{6.6}cm^3} = 3.3g/cm^3}$$

d. 2.6 × 10^5 + 4.1 × 10^7

2.6 × 10^5 + 410 × 10^7 = 410 × 10^5 = 4.1 × 10^7

e. $\dfrac{\left(6.02 \times 10^2\right) \times 0.32000}{\left(5.325 + 21.0\right) \times \left(45.44 + 0.32\right)}$

$$\mathbf{\frac{\left(\overset{3}{6.02} \times 10^2\right) \times \overset{5}{0.32000}}{\left(5.325 + 21.0\right) \times \left(45.44 + 0.32\right)} = \frac{\overset{3}{193}}{\underset{3}{26.3} \times \underset{4}{45.76}} = \frac{\overset{3}{193}}{\underset{3}{1200}} = 0.161}$$

12. How many significant figures are in each of the following numbers?
   a. 5.40 km = **3**
   b. 210 μL = **2**
   c. 801.5 g = **4**
   d. 1,000 ns = **1**
   e. 101.0100 mL = 7
   f. 1.2 × 10^3 mm = **2**
   g. 0.00120 cg
   h. 0.0102 ds = **3**
   i. 9.010 × 10^{-6} ks = **4**
   j. 2,370.0 μg = **5**

13. Round off each of the following numbers to four significant figures:
   a. 15.9994 nL = **16.0 nL**
   b. 0.66549 s = **0.665 s**
   c. 87,550 mg = **87,600 mg**
   d. 1.0080 kL = **1.01 kL**
   e. 4885 μm = **4890 μm**
   f. 0.027225 s = **0.0272 s**

14. Perform the following mathematical operations and express your answers to the proper number of significant figures and units:
   a. 642 m × (4.0 × 10^{-5} m) = **2.6 × 10^{-2} m^2**
   b. 17 g ÷ (3.88 × 10^7 cm^3) = **4.4 × 10^{-7} g/cm^3**
   c. (2.9 × 10^{-5} cg) + (8.1 × 10^{-4} cg) = **8.4 × 10^{-4} cg**
   d. (4.3 × 10^{-5} cm)^3 = **8.0 × 10^{-14} cm^3**
   e. (5.40 × 10^{-18} kg) / 769 mL = **7.02 × 10^{-21} kg/mL**
   f. [59 cm × (3.24 × 10^{-2} cm)] ÷ (4.80 × 10^4 cm) = **4.0 × 10^{-5} cm**
   g. [42 dm × (4.02 × 10^{23} dm)] ÷ 0.016 dm = **1.1 × 10^{27} dm**
   h. 123.040 kg – 55.06 kg = **67.98 kg**
   i. 0.00000016 g / 74.3 L = **2.2 × 10^{-9} g/L**
   j. 10.0 ns + 54.600 ns = **64.6 ns**

15. Using two different instruments, I measured the length of my foot to be 27 centimeters and 27.00 centimeters. Explain the difference between these two measurements.

    The two numbers imply different precision. The 27 centimeters is not as precise a measurement as the 27.00 centimeters.

16. Determine the number of significant figures in each of the following numbers:
    a. 23,450,000 μg = **4**
    b. 0.00340 kL = **3**
    c. 24.90000 dm = **7**
    d. 32,001,000 cs = **5**
    e. 3.4500×10^3 ng = **5**
    f. 0.102 mL = **3**

17. Calculate the following, answer using proper significant figures and units:
    a. 43.00 dL + 19.1 dL = **62.1 dL**
    b. 6.500 ms + 120 ms = **130 ms**
    c. (13.980 cg – 11.0 cg) + (340 cg + 217 cg) = **560 cg**
    d. 2.7800×10^4 nm + 1.10×10^4 nm = **3.88×10^4 nm**
    e. 0.00450 mm × 3.0454 mm × 1.0000 mm = **0.0137 mm³**
    f. $\frac{0.0320m \times 9.100m \times 0.0045m}{4320m}$ = **3.0×10^{-7} m²**
    g. $\frac{0.00321g}{0.02000mL} \times 2390mL$ = **384 g**
    h. $\frac{(4230g + 547g)}{0.1340cm \times 0.050cm \times 3.10cm}$ = **2.3×10^5 g/cm³**
    i. $(8.900 \times 10^8$ km$) \times (3.20 \times 10^{-5}$ km$)$ = **2.85×10^4 km²**

TEMPERATURE

1. Convert 25.0°C to
    a. **°F = 1.8°C + 32°F = (1.8 × 25.0) + 32 = 77.0°F**
    b. **K = 273.1 + °C = 273.1 + 25.0 = 298.1 K**

2. Convert 345 K to
    a. **°F = (Do b first) = 1.8°C + 32°F = (1.8 × 72) + 32**
    b. **°C = K – 273 = 345 – 273 = 72°C**

3. Convert the following temperatures:
    a. 250 Kelvin to Celsius = **–20°C**
    b. 17 Celsius to Kelvin = **290. K**
    c. 89.5 Fahrenheit to Celsius = **31.9°C**
    d. 339 Kelvin to Celsius = **66°C**
    e. 55 Celsius to Kelvin = **328 K**
    f. 383 Kelvin to Fahrenheit = **230.°F**

4. Convert the following temperatures:
   a. 32°F to °C = **0°C**
   b. 988°C to °F = **1810. K**
   c. -65.8°F to °C to K = **–54.3°C, 218.8 K**
   d. 243.8 K to °C to °F = **–29.4°C, –20.8°F**

5. What is the base unit for temperature?

   Kelvin

6. What is the temperature in degree Celsius of 452 K?

   179°C

7. What is the temperature in Kelvin of 67°F?

   293 K

VOLUME AND DENSITY

1. Calculate the volume of a piece of metal that has a length of 4.5 cm, a height of 3.6 cm, and a width of 8.9 cm.

   V = l × w × h
   V = 4.5 cm × 3.6 cm × 8.9 cm = 144.18 cm^3 = 140 cm^3

2. If a piece of copper has a volume of 51 cm^3 and a mass of 457.00 grams, what is its density in g/mL?

   $$D = \frac{m}{V} = \frac{457.00g}{51cm^3} = 8.96g/cm^3 = 9.0g/cm^3$$

3. Calculate the volume (in mL) of a metal cube with the following dimensions:
   length = 4.5 cm, width = 6.7 cm, height = 2.4 cm

   72 cm^3

4. Calculate the volume (in cm^3) of a metal cylinder with the following dimensions:
   radius = 93.4 cm, height = 67800 cm

   1.86 × 10^9 cm^3

5. Calculate the volume of a metal cube (in cm^3) with the following dimensions:
   length = 0.98 dm, width = 990 mm, height = 3.3 cm

   3200 cm^3

6. 56.7 mL of water was added to a 100.0 mL graduated cylinder. A solid sample was placed into the graduated cylinder, completely submerged in the water. The water level rose to 89.4 mL. What is the volume of the solid sample (in cm^3)?

   32.7 cm^3

7. Calculate the density (g/mL) of a solid sample having a mass of 367 grams and a volume of 89.4 cm^3.
   4.11 g/mL

8. Calculate the volume (in L) of a solid sample having a mass of 57.8 grams and a density of 1.26 g/cm^3.
   0.0459 Lb

9. The density of platinum metal is 21.5 g/cm^3. Calculate the mass (in g) of a block of platinum that has a volume of 0.876 L.
   18,800 g

10. A lead block has a length of 0.22 meters, a width of 365 millimeters, a height of 4.8 decimeters, and a mass of 439 kilograms. Based on this information, what is the density of lead in g/cm^3?
    11 g/cm^3

11. If the density of alcohol is 0.789 g/mL, how many mL of alcohol must be measured out for an experiment that requires 2.50 grams of alcohol?
    3.17 mL

12. Suppose you wish to purchase a waterbed that has the dimensions 2.45 m × 21.5 dm × 23 cm. How many kilograms of water does this bed contain? (The density of water is 1.00 g/cm^3.)
    1,200 kg

13. The mass of a toy spoon is 7.5 grams, and its volume is 3.2 mL. What is the density of the toy spoon?
    2.3 g/mL

14. A mechanical pencil had the density of 3.0 g/cm^3. When the pencil was placed into 56.8 mL of water, the water level rose to 73.1 mL (the pencil being completely submersed).
    a. What is the volume of the pencil?
       16.3 mL

    b. What is the mass of the pencil?
       49 g

15. A screwdriver has the density of 5.5 g/mL. The mass of the screwdriver is 2.3 grams. What is the screwdriver's volume?
    0.42 mL

16. A metal cylinder has a mass of 5.48 grams. The measured height is 5.6 cm and diameter is 445 dm.
    a. What is the volume of the metal cylinder?
       8.7×10^7 cm^3

    b. What is the density of the metal cylinder?
       6.3×10^{-8} g/cm^3

17. What is the volume of a metal cube with the following dimensions?
    length = 3.2 cm, width = 2.1 cm, height = 1.5 cm
    volume = length × width × height

    10.1 cm^3

18. What is the definition of density?

    mass/volume

19. What is the density of a metal if a 15.4 gram sample has a volume of 1.96 cm^3?

    7.86 g/cm^3

20. 5.00 mL = ____________ cm^3?

    5.00 cm^3

21. 2.5 dL = ____________ cm^3?

    250 cm^3

22. What is the volume of a rectangular solid having the following dimensions?
    length = 8.30 cm, width = 3450 mm, height = 0.0540 m

    15500 cm^3

23. What volume of a liquid having a density of 3.48 g/cm^3 is needed to supply 5.00 grams of the liquid?

    1.44 cm^3

24. A metal sample having a mass of 30.9 grams was added to a graduated cylinder containing 23.2 mL of water. The volume of the water plus the sample was 24.8 mL. What is the density of the metal?

    19.3 g/mL

25. An unknown liquid has a density of 0.786 liters and a density of 19.4 g/cm^3. What is the mass of the liquid?

    15,200 g

UNIT 3: PRACTICE PROBLEMS

HISTORY OF THE ATOM

What does the atom look like?

Early Discoveries:

J. J. Thomson   Ernest Rutherford   Democritus   John Dalton   James Chadwick   Robert Millikan

Marie Curie   Dmitri Mendeleev

Gold foil experiment   Alpha particles through nitrogen   Cathode ray tube   Plum Pudding Model

Oil Drop experiment   Discovered radium   Arranged Elements based upon Atomic Weight (Mass)

The Greek's Atom, ~400 BC © Georgios Kollidas/ Shutterstock.com **Democritus**	**The Atomic Theory, 1803** © Georgios Kollidas/ Shutterstock.com **John Dalton**
Periodic Table, 1869 © Olga Popova/ Shutterstock.com **Dmitri Mendeleev, Arranged Elements based upon Atomic Weight (Mass)**	**The Electron, 1897** © felichy/ Shutterstock.com **J.J. Thomson, Cathode ray tube**
Radioactive Elements, 1898 © catwalker/ Shutterstock.com **Marie Curie, Discovered radium**	**The Electron, 1909** © catwalker/ Shutterstock.com **Robert Millikan, Oil Drop experiment**
The Proton, 1902 © rook76/ Shutterstock.com **Ernest Rutherford, Alpha particles through nitrogen**	**The Neutron, 1932** © general-fmv/ Shutterstock.com **James Chadwick**
The Atom, 1904 © Iaryna Turchyniak/ Shutterstock.com **J.J. Thomson, Plum pudding model**	**The Atom, 1911** © Iaryna Turchyniak/ Shutterstock.com **Ernest Rutherford, Gold foil experiment**

ISOTOPES

Two Atoms of Lithium:

Lithium-6 has $3p^+$, $3n^0$, $3e^-$	Lithium-7 has $3p^+$, $4n^0$, $3e^-$

Element	Number of Protons	Number of Neutrons	Number of Electrons	Atomic Number	Mass Number	Average Atomic Mass
lithium	**3**	4	**3**	**3**	**7**	**6.941 amu**
carbon	**6**	7	**6**	**6**	13	**12.011 amu**
silver	47	61	**47**	**47**	**108**	**107.868 amu**
lead	**82**	**125**	**82**	**82**	207	**207.2 amu**
calcium	20	19	**20**	**20**	**39**	**40.078 amu**
tantalum	**73**	**108**	**73**	**73**	181	**180.948 amu**
radium	88	**137**	**88**	**88**	225	**(206) amu**
samarium	**62**	**89**	**62**	**62**	151	**150.36 amu**

Creating a Cation:

Neutral atom of Magnesium: Mg has $12p^+$, $12e^-$	Charged atom of Magnesium: Mg^{2+} has $12p^+$, $10e^-$

Creating an Anion:

Neutral atom of Nitrogen: N has $7p^+$, $7e^-$	Charged atom of Nitrogen: N^{3-} has $7p^+$, $10e^-$

QUESTION GROUP #1

Directions and/or Common Information:

Isotope Name	Isotope Mass (amu)	Percentage
Silver-107	107	51.86
Silver-109	109	remainder

Find the missing percentage.

1. 48.14%

Find the average atomic mass of an atom of silver.

2. 107.963 amu

QUESTION GROUP #2

Directions and/or Common Information: Silicon has three naturally occurring isotopes.

Isotope Name	Isotope Mass (amu)	Percentage
Silicon-28	28	92.21
Silicon-29	29	4.70
Silicon-30	30	3.09

Now, find the average atomic mass of an atom of silicon

1. 28.109 amu

QUESTION GROUP #3

Directions and/or Common Information: Iron has four isotopes.

Isotope Name	Isotope Mass (amu)	Percentage
Iron-54	5.90%	54
Iron-56	91.72%	56
Iron-57	2.10%	57
Iron-58	0.280%	58

Estimate the average mass.

1. 55

Calculate the average atomic mass of iron.

2. 55.909

M&M® CANDY ISOTOPE ACTIVITY

This is Isotopes for the Element: Mm

Red Isotope	Blue Isotope	Yellow Isotope	Green Isotope	Brown Isotope	Orange Isotope	
Ave. Mass (Red) 52.00 amu	**Ave. Mass (Blue) 53.00 amu**	**Ave. Mass (Yellow) 55.00 amu**	**Ave. Mass (Green) 56.00 amu**	**Ave. Mass (Brown) 54.00 amu**	**Ave. Mass (Orange) 57.00 amu**	
Amount (red) = **5**	Amount (blue) = **12**	Amount (yellow) = 7	Amount (green) = **10**	Amount (brown) = **3**	Amount (orange) = 6	**TOTAL AMOUNT OF M&Ms (in package)**
Relative abundance (red) = **0.1163**	Relative abundance (blue) = **0.2791**	Relative abundance (yellow) = **0.1628**	Relative abundance (green) = **0.2326**	Relative abundance (brown) = **0.06977**	Relative abundance (orange) = **0.1395**	**RELATIVE ABUNDANCE = amount/total**
Percent abundance (red) = **11.63%**	Percent abundance (blue) = **27.91%**	Percent abundance (yellow) = **16.28%**	Percent abundance (green) = **23.26%**	Percent abundance (brown) = **6.977%**	Percent abundance (orange) = **13.95%**	**PERCENT ABUNDANCE = rel. abundance × 100**
Relative mass (red) = **6.048 amu**	Relative mass (blue) = **14.79 amu**	Relative mass (yellow) = **8.954 amu**	Relative mass (green) = **13.03 amu**	Relative mass (brown) = **3.769 amu**	Relative mass (orange) = 7.952	**RELATIVE MASS = rel. abundance × ave. mass**

Calculate the Average Atomic Mass of the Element: Mm = **54.54 amu**

Average atomic mass = total of relative masses

PERIODIC TABLE OF THE ELEMENTS

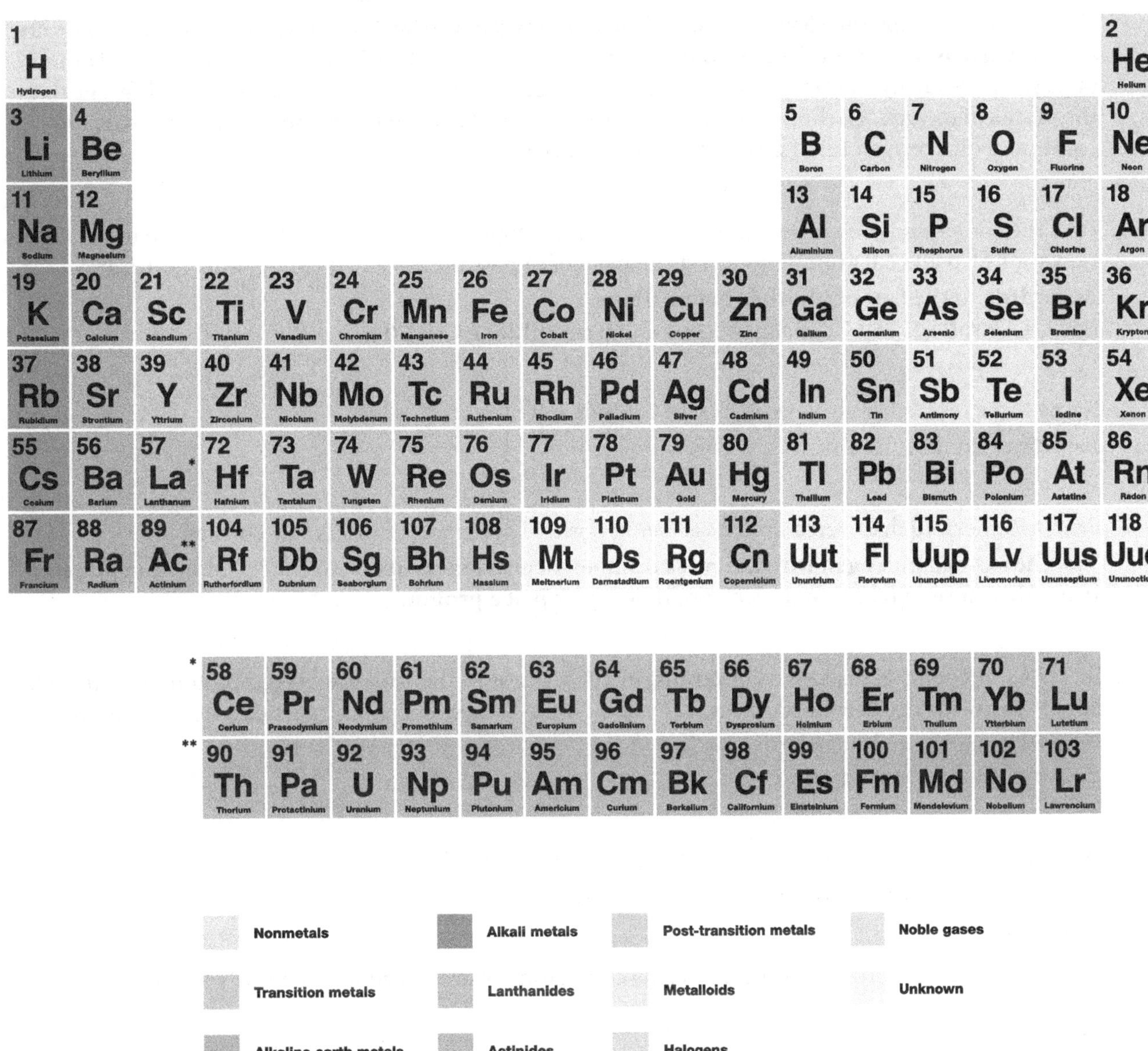

HISTORY OF THE ATOM

1. How did Dalton come up with the Atomic Theory?

   By very little, if any experimentation. Dalton made observations during his weather studies, read of past discussions on the atom (including Democritus and Newton), and started making predictions about atoms being in existence. After his theory was put out into the scientific community, others agreed and disputed it. Today we accept the Atomic Theory because others to follow Dalton found enough evidence to support the theory.

2. Discuss the two compounds (I) carbon monoxide (CO) and (II) carbon dioxide (CO_2). Using Dalton's atomic theory, explain how these two compounds (and their elements) exist.

   Discuss with group: The elements are carbon and oxygen. Carbon and oxygen atoms can not be created or destroyed to form the two compounds. All the atoms of carbon are identical and all atoms of oxygen are identical (except for the amount of neutrons). The atoms of carbon are different than the atoms of oxygen. Carbon combines with oxygen in a 1:1 ratio for carbon monoxide, carbon combines with oxygen in a 1:2 ratio for carbon dioxide.

3. Take the compounds (I) copper oxide (Cu_2O) and (II) copper oxide: CuO. If the mass ratio in Cu_2O is 12.0 grams Cu and 6.0 grams O; suggest the mass of O possible in CuO if Cu remains 12.0 grams.

   According to the Law of Multiple Proportions:
   For Cu_2O: ratio 12.0 g Cu: 6.0 g O For CuO: ratio 12.0 g Cu: 3.0 g O

4. Name three researchers who worked with cathode ray tubes.

   Faraday, Crookes, Thomson

5. Ernest Rutherford discovered protons as being positive hydrogen atoms. Is this correct?

   Yes, the isotope hydrogen with the mass number 1 contains no neutrons, one proton, and one electron. Therefore, if it loses one electron, it becomes just a proton.

6. In 1897, J. J. Thomson discovered electrons, in 1911 Ernest Rutherford discovered protons, and in 1932 James Chadwick discovered neutrons. Why do you think these subatomic particles were discovered in this order?

   Your opinion from your readings.

7. Why is Rutherford's model of the atom better than Thomson's?

   It included a nucleus containing protons and neutrons.

8. Who helped Rutherford discover the nucleus of an atom by an experiment famously known as the gold foil experiment?

   Marsden & Geiger

9. Which one of Dalton's major conclusions explained the law of conservation of mass?
   a. Matter is composed of small, indivisible particles called atoms.
   b. Atoms of the same element are identical and have the same properties.
   c. Compounds are composed of atoms of different elements combined in small whole-number ratios.
   d. Atoms of the same element have the same mass.
   e. Reactions are merely the rearrangement of atoms into different combinations.

10. Who contributed the following: The charge of the electron was determined by performing the oil drop experiment.

    Millikan

11. Who contributed to the discovery of the neutron?

    Chadwick

12. Who contributed the following: The charge/mass ratio for 20 different metals in the cathode ray tube was the same, which suggested that electrons are present in all kinds of matter.

    Thomson

13. Taken by itself the fact that 8.0 grams of oxygen and 1.0 grams of hydrogen combine to give 9.0 grams of water demonstrate what natural law?
    a. conservation of energy
    b. conservation of mass
    c. periodicity
    d. atomic theory
    e. atomic number

14. Who discovered the nucleus in the atom?

    Rutherford

STRUCTURE OF THE ATOM—PROTONS, NEUTRONS, ELECTRONS

1. The mass number of an atom is equal to ______.
   a. the number of electrons in the nucleus
   b. the number of protons in the nucleus
   c. the number of neutrons in the nucleus
   d. the sum of the protons and neutrons in the nucleus

2. The number of protons found in an atom is equal to ______.
   a. the number of electrons in the nucleus
   b. the atomic number of the element
   c. the number of neutrons in the nucleus
   d. the sum of the protons and neutrons in the nucleus

3. The number of neutrons found in an atom is equal to ______.
   a. the number of electrons in the nucleus
   b. the mass number of the atom
   c. the number of protons in the nucleus
   d. the difference between the mass number and the atomic number in the atom

4. Complete the following table.

Atom	Charge	Ion	#p+	#e-
Sr	+2	Sr^{2+}	38	38–2 = 36
O	**–2**	O^{2-}	**8**	**8+2 = 10**
C	+4	**C^{4+}**	6	**6–4 = 2**
Fe	+3	**Fe^{3+}**	**26**	**26–3 = 23**
Sb	–3	**Sb^{3-}**	**51**	**51+3 = 54**
Br	**–1**	Br^{-}	**35**	**35+1 = 36**

5. How many protons, neutrons, and electrons does the element $^{65}_{30}Zn^{2+}$ contains:

   30 protons, 35 neutrons, 28 electrons

6. What is the total number of subatomic particles in ${}^{44}_{21}Sc^{3+}$?

62

7. Which of the following is an isotope of ${}^{204}_{82}Pb$?
   a. ${}^{199}_{81}Tl$
   b. ${}^{202}_{80}Hg$
   c. ${}^{212}_{82}Pb$
   d. ${}^{206}_{80}Hg$
   e. ${}^{197}_{81}Tl$

8. The isotopes of the same element have ______.
   a. different number of protons
   b. different number of neutrons
   c. different atomic numbers
   d. different number of electrons

9. Find the number of the neutrons and electrons for lead-208 (this is the mass number).

126 neutrons, 82 electrons

10. Complete the following table.

Element	Atomic Number	Mass Number	#p⁺	#n⁰	#e⁻	Isotopic Notation
K	19	39	19	20	19	${}^{39}_{19}K$
Br	**35**	80	**35**	**45**	**35**	**${}^{80}_{35}Br$**
Ar	**18**	**40**	**18**	22	**18**	**${}^{40}_{18}Ar$**
Cr	**24**	52	**24**	**28**	22	**${}^{52}_{24}Cr^{2+}$**
Si	**14**	**28**	**14**	14	10	**${}^{28}_{14}Si^{4+}$**

11. Which one of the following statements about the isotopes of a given element is correct?
   a. The atoms have the same number of protons but differing numbers of neutrons.
   b. The atoms have the same number of neutrons but differing numbers of electrons.
   c. The atoms have the same number of electrons but differing numbers of protons.
   d. The atoms have the same number of neutrons but differing numbers of protons.
   e. The atoms have the same number of protons but differing numbers of electrons.

UNIT 4: PRACTICE PROBLEMS

IONIC NOMENCLATURE

Ionic bond (form **salts, particles, or formula units**):

Ionic bonds are bonds between: **metal + nonmetal (or polyatomic)** Exception: **ammonium (NH_4^+)**

KBr	**potassium bromide**
$BaSO_4$	**barium sulfate**
Cu_2O	**copper (I) oxide** **cuprous oxide**
$Fe_2(SO_4)_3$	**iron (III) sulfate** **ferric sulfate**
K_3PO_4	**potassium phosphate**
Al_2O_3	**aluminum oxide**
NH_4Cl	**ammonium chloride**
CuOH	**copper (I) hydroxide** **cuprous hydroxide**
$Ca(C_2H_3O_2)_2$	**calcium acetate**
$ZnSO_3$	**zinc sulfite**
$SnCl_2$	**tin (II) chloride** **stannous chloride**

Barium nitrate	$Ba(NO_3)_2$
Potassium iodide	KI
Iron (III) bromide	$FeBr_3$
Copper (I) phosphate	Cu_3PO_4
Zinc sulfite	$ZnSO_3$
Calcium hydroxide	$Ca(OH)_2$
Tin (IV) sulfide	SnS_2
Magnesium carbonate	$MgCO_3$
Ammonium nitrate	NH_4NO_3
Aluminum sulfate	$Al_2(SO_4)_3$
Tin (II) bromide	$SnBr_2$
Sodium bicarbonate	$NaHCO_3$

ammonium sulfide **$(NH_4)_2S$**	nickel (II) iodide **NiI_2**	sodium nitrate **$NaNO_3$**	mercurous oxide **Hg_2O**
cupric bromide **$CuBr_2$**	lead(II) chlorite **$Pb(ClO_2)_2$**	aluminum sulfate **$Al_2(SO_4)_3$**	potassium nitrate **KNO_3**
iron(II) bisulfate **$Fe(HSO_4)_2$**	ferrous carbonate **$FeCO_3$**	magnesium nitrate **$Mg(NO_3)_2$**	lead(II) phosphate **$Pb_3(PO_4)_2$**
iron(III) chromate **$Fe_2(CrO_4)_3$**	iron(II) chromate **$FeCrO_4$**	cupric hydroxide **$Cu(OH)_2$**	copper(II) hydroxide **$Cu(OH)_2$**
calcium fluoride **CaF_2**	cuprous carbonate **Cu_2CO_3**	nickel (II) nitrate **$Ni(NO_3)_2$**	chromic acetate **$Cr(C_2H_3O_2)_3$**
silver cyanide **AgCN**	calcium chlorate **$Ca(ClO_3)_2$**	$(NH_4)_2SO_3$ **ammonium sulfite**	$(NH_4)_2O$ **ammonium oxide**
$ZnSO_4$ **zinc sulfate**	$Al(ClO_4)_3$ **aluminum perchlorate**	$SnCl_2$ **tin(II) chloride**	$Zn(HCO_3)_2$ **zinc bicarbonate**
$FeCl_3$ **iron(III) chloride**	Na_3PO_4 **sodium phosphate**	Ag_2S **silver sulfide**	AgClO **silver hypochlorite**
$Mg(OH)_2$ **magnesium hydroxide**	$(NH_4)_3PO_4$ **ammonium phosphate**	$(NH_4)_2CO_3$ **ammonium carbonate**	$Fe(ClO_2)_2$ **ferrous chlorite**
$Ni(C_2H_3O_2)_2$ **nickel(II) acetate**	K_2S **potassium sulfide**	Na_2CrO_4 **sodium chromate**	$SnBr_4$ **tin(IV) bromide**
$Cr(HSO_4)_3$ **chromic bisulfate**	Li_2CrO_4 **lithium chromate**	$KMnO_4$ **potassium permanganate**	$Mg(HSO_4)_2$ **magnesium bisulfate**
$AgClO_4$ **silver perchlorate**	$Fe_3(PO_4)_2$ **ferrous phosphate**	K_3PO_4 **potassium phosphate**	

Molecular Nomenclature

nitrogen trihydride **NH_3**	phosphorus pentachloride **PCl_5**	phosphorus triiodide **PI_3**
arsenic pentachloride **$AsCl_5$**	diphosphorus trioxide **P_2O_3**	nitrogen dioxide **NO_2**
dinitrogen tetrafluoride **N_2F_4**	pentaboron nonahydride **B_5H_9**	tetrasulfur tetranitride **S_4N_4**
H_2O_2 **dihydrogen dioxide** **hydrogen peroxide**	CS_2 **carbon disulfide**	BCl_3 **boron trichloride**
SCl **sulfur monochloride**	CCl_4 **carbon tetrachloride**	S_2Cl_2 **disulfur dichloride**
NO **nitrogen monoxide** **nitrous oxide**	OF_2 **oxygen difluoride**	P_4S_{10} **tetraphosphorus decasulfide**

Acid Nomenclature

hydroiodic acid **HI**	chlorous acid **$HClO_2$**	chloric acid **$HClO_3$**
sulfurous acid **H_2SO_3**	hydrosulfuric acid **H_2S**	hydroselenic acid **H_2Se**
selenic acid **H_2SeO_4**	periodic acid **HIO_4**	oxalic acid **$H_2C_2O_4$**
HF **hydrofluoric acid**	H_2CO_3 **carbonic acid**	HNO_2 **nitrous acid**
H_6TeO_6 **tilluric acid**	$H_2Cr_2O_7$ **dichromic acid**	HCl **hydrochloric acid**
HOCN **cyanic acid**	$HBrO_2$ **bromous acid**	$HC_2H_3O_2$ **acetic acid**

Combined Nomenclature

silver bicarbonate **$AgHCO_3$**	nitrous acid **HNO_2**	barium bisulfate **$Ba(HSO_4)_2$**
lead(IV) chlorite **$Pb(ClO_2)_4$**	hydrobromic acid **HBr**	calcium sulfide **CaS**
triphosphorus heptoxide **P_3O_7**	hypochlorous acid **HClO**	copper(I) bisulfate **$CuHSO_4$**
sulfuric acid **H_2SO_4**	boron triphosphide **BP_3**	cobaltic chlorate **$Co(ClO_3)_3$**
SO_3 **sulfur trioxide**	$HClO_4$ **perchloric acid**	NH_3 **nitrogen trihydride**
$Ba(BrO_3)_2$ **barium bromate**	XeF_2 **xenon difluoride**	Al_2S_3 **aluminum sulfide**
H_3PO_4 **phosphoric acid**	$Mn_3(PO_4)_2$ **manganese (II) phosphate**	H_2Se **hydroselenic acid**
H_2SO_3 **sulfurous acid**	P_2O_4 **diphosphorus tetroxide**	HNO_3 **nitric acid**
$Cr_2(SO_3)_3$ **chromic sulfite**	$Ni(ClO_4)_2$ **nickel (II) perchlorate**	H_3AsO_4 **arsenic acid**
sodium acetate **$NaC_2H_3O_2$**	carbonic acid **H_2CO_3**	tetrasulfur hexaphosphide **S_4P_6**

STRUCTURE OF CHEMICAL COMPOUNDS

1. C_8H_{18} contains a total of __26__ atoms, including __8__ atoms C and __18__ atoms H.
2. $Al(NO_3)_3$ contains a total of 13 atoms, including __3__ atoms N, __1__ atoms Al, and __9__ atoms O.
3. $(NH_4)_2SO_4$ contains a total of __15__ atoms.

4. One particle of $(NH_4)_3PO_4$, ammonium phosphate, contains **3** NH_4^+ ions, **1** PO_4^{3+} ions, **3** atoms N, **12** atoms H, **1** atoms P, **4** atoms O, and **20** total atoms.

5. Which of the following is not a diatomic molecule?
   a. H_2
   b. N_2
   c. K_2
   d. O_2
   e. Cl_2

6. The total number of atoms represented by $Ba(H_2PO_4)_2$ is:
   a. 16
   b. 13
   c. 14
   d. 17
   e. 15

7. Classify each compound as either: M = molecular (two nonmetals) I = Ionic (metal + nonmetal)
   a. $FeCl_2$ **Ionic**
   b. N_2O **Molecular**
   c. HBr **Molecular**
   d. CO_2 **Molecular**
   e. BaI_2 **Ionic**
   f. K_2O **Ionic**

8. List 3 common examples of each of the three types of bonding.
   a. Ionic **table salt: NaCl, baking soda: $NaHCO_3$, rust: Fe_2O_3**
   b. Covalent **water: H_2O, carbon dioxide: CO_2, carbon monoxide: CO**
   c. Metallic **gold: Au, silver: Ag, platinum: Pt**

9. Circle each of the following compounds that is ionically bonded:
   CO_2 SiO_2 (**NaCl**) HBr CH_4 (**Rb_2O**) (**KF**)

10. Which of the following compounds are covalently bonded?
   (**NO_2**) (**PCl_3**) (**NH_3**) NaI $CaCl_2$ Al_2O_3 (**C_6H_6**)

11. Which of the following pairs of atoms would not be expected to combine to form an ionic compound?
   a. B and H
   b. Na and H
   c. Li and Cl
   d. Rb and O
   e. K and S

12. Which of the following two elements would be expected to form a covalent bond?
   a. Na, O
   b. H, S
   c. K, Te
   d. Mn, F
   e. Ba, S

13. Which of the following is a covalent compound?
    a. SCl
    b. FrCl
    c. CaS
    d. BaS
    e. KCl

14. Which of the following is an ionic compound?
    a. NO
    b. CO
    c. OF_2
    d. O_2
    e. K_2O

15. Which of the following is an ionic compound?
    a. H_2O
    b. CS_2
    c. $CaCl_2$
    d. CO_2
    e. CH_4

16. Which of the following is a covalent compound?
    a. NaCl
    b. $FeCl_3$
    c. $BaCl_2$
    d. NCl_3
    e. $AlCl_3$

17. What do you call the following compound: Na_3PO_4?
    a. a molecule
    b. a polyatomic ion
    c. a particle
    d. a molecule or a particle

IONS: IDENTIFYING CATIONS AND ANIONS

1. What is the expected charge on the Ba ion?

   +2

2. What is the expected charge on the alkaline earth metals?

   +2

3. Which of the following forms a +3 ion?
   a. Na
   b. Mg
   c. Al
   d. Si
   e. P

4. Complete the following table:

Element	Ion(s)	Oxidation Number(s)	Names
Iron	Fe^{2+}	+2	iron (II) or ferrous
	Fe^{3+}	+3	iron (III) or ferric
Potassium	**K^{+}**	+1	**potassium**
Beryllium	**Be^{2+}**	+2	**beryllium**
Tin	**Sn^{2+}**	+2	**tin (II) or stannous**
	Sn^{4+}	+4	**tin (IV) or stannic**
Cobalt	**Co^{2+}**	+2	**cobalt (II)**
	Co^{3+}	+3	**cobalt (III)**
Oxygen	**O^{2-}**	**−2**	**oxide**
Fluorine	**F^{-}**	**−1**	**fluoride**
Copper	**Cu^{+}**	+1	**copper (I) or cuprous**
	Cu2+	+2	**copper (II) or cupric**

5. Give the ions present and their relative numbers in K_2SO_4.
   a. 1 K^+ and 1 SO_4^-
   b. 2 K^+ and 1 SO_4^{2-}
   c. 1 K^+ and 2 SO_4^{2-}
   d. 2 K^+ and 2 SO_4^{2-}
   e. 2 K^{2+} and 1 SO_4^{2-}

6. What is the symbol for the chlorite ion?
   a. ClO^-
   b. CrO_4^-
   c. ClO_2^-
   d. ClO_3^-
   e. ClO_4^-

7. What is the symbol for the sulfite ion?
   a. SO_3^-
   b. SO_3^{2-}
   c. SO_4^{2-}
   d. SO_2^{2-}
   e. SO_4^-

8. What is the symbol for the chlorate ion?
   a. CrO_4^-
   b. ClO_2^{3-}
   c. ClO^-
   d. ClO_3^-
   e. ClO_4^-

NOMENCLATURE: WRITING FORMULAS & NAMING CHEMICAL COMPOUNDS
IONIC

1. Given the following ions, write the correct formula: Ba^{2+} and PO_4^{3-}
   a. $BaPO_4$
   b. $Ba_2(PO_4)_3$
   c. Ba_2PO_{12}
   d. Ba_3PO_8
   e. $Ba_3(PO_4)_2$

2. What is the chemical formula for lead (IV) sulfide?
   a. PbS_4
   b. PbS
   c. Pb_4S
   d. PbS_2
   e. Pb_2S

3. What is the chemical name of $Fe(HSO_3)_3$?
   a. iron (III) sulfite
   b. iron (III) bisulfite
   c. iron (I) bisulfite
   d. iron (III) sulfate
   e. iron (III) bisulfate

4. What is the chemical name of $BaSO_4$
   a. Barium sulfur
   b. Barium sulfoxide
   c. Barium sulfide
   d. Barium sulfite
   e. Barium sulfate

5. What is the chemical name of $Mg_3(PO_4)_2$?
   a. Magnesium phosphide
   b. Magnesium phosphotate
   c. Magnesium phosphate
   d. Magnesium (II) phosphate

6. Write the chemical formula for each of the following compounds:

potassium sulfate	**K_2SO_4**	sodium oxide	**Na_2O**
aluminum nitride	**AlN**	iron(II) perchlorate	**$Fe(ClO_4)_2$**
calcium carbonate	**$CaCO_3$**	chromium (III) chloride	**$CrCl_3$**
cesium bromide	**CsBr**	barium sulfate	**$BaSO_4$**
beryllium oxide	**BeO**	sodium bicarbonate	**$NaHCO_3$**
tin (II) fluoride	**SnF_2**	cobalt (III) nitrate	**$Co(NO_3)_3$**
potassium chlorate	**$KClO_3$**	lithium hydroxide	**LiOH**
copper (I) sulfide	**Cu_2S**	calcium iodide	**CaI_2**

7. Write the chemical name for each of the following compounds

Cu_2S	**copper (I) sulfide**	SnO_2	**tin (IV) oxide**
CrO_3	**chromium (VI) oxide**	Cr_2O_3	**chromium (III) oxide**
Al_2O_3	**aluminum oxide**	NaCl	**sodium chloride**
$MgCl_2$	**magnesium chloride**	RbBr	**rubidium bromide**
CsF	**cesium fluoride**	AlI_3	**aluminum iodide**
$FePO_4$	**iron (III) phosphate**	$NaHSO_4$	**sodium hydrogen sulfate**
KNO_3	**potassium nitrate**	$PtCl_4$	**platinum (IV) chloride**
NH_4C	**ammonium chloride**		

NOMENCLATURE: WRITING FORMULAS & NAMING CHEMICAL COMPOUNDS MOLECULAR (COVALENT)

1. What is the correct name for N_2S?
   a. nitrogen monosulfide
   b. mononitrogen disulfide
   c. nitrogen sulfur
   d. dinitrogen disulfide
   e. dinitrogen monosulfide

2. The formula of the compound trichlorine tetrafluoride is _____.
   a. ClF
   b. Cl_3F_4
   c. Cl_3F_5
   d. Cl_5F_3
   e. Cl_4F_3

3. Write the chemical formula for each of the following compounds:

diphosphorus trioxide	**P_2O_3**	arsenic pentachloride	**$AsCl_5$**
dihydrogen monoxide	**H_2O**	chlorine monoxide	**ClO**
sulfur difluoride	**SF_2**	oxygen difluoride	**OF_2**
sulfur dioxide	**SO_2**		

4. Write the chemical name for each of the following compounds

H_2	**hydrogen**	H_2O	**water or dihydrogen monoxide**
CO_2	**carbon dioxide**	NH_3	**ammonia or nitrogen trihydride**
SO_3	**sulfur trioxide**	CH_4	**methane or carbon tetrahydride**
P_2S_6	**diphosphorus hexasulfide**	N_2O_4	**dinitrogen tetroxide**
C_2F_2	**dicarbon difluoride**	SeO_2	**selenium dioxide**
SeO_3	**selenium trioxide**	NI_3	**nitrogen triiodide**
PCl_3	**phosphorous trichloride**	SF_2	**sulfur difluoride**
NO	**nitrogen monoxide**	NF_3	**nitrogen trifluoride**
N_2F_4	**dinitrogen tetrafluoride**	H_2SO_4	**sulfuric acid**
SiF_4	**silicon tetrafluoride**		
NH_4Cl	**ammonium chloride**		

NOMENCLATURE: WRITING FORMULAS & NAMING CHEMICAL COMPOUNDS ACIDS

1. What is the correct name for HCl?
   a. hydrogen sulfur
   b. chloric acid
   c. chlorous acid
   d. hydrochloric acid
   e. hydrogen chlorine

2. What is the name of H_2S?
   a. hydrogen sulfate
   b. sulfuric acid
   c. sulfurous acid
   d. hydrosulfuric acid

3. Write the chemical formula for each of the following compounds:
   a. carbonic acid **H_2CO_3**
   b. hydrobromic acid **HBr**
   c. hypochlorous acid **HClO**
   d. hydrochloric acid **HCl**
   e. nitrous acid **HNO_2**

4. Write the chemical name for each of the following compounds:
   a. HNO_2 **nitrous acid**
   b. $HBrO_2$ **bromous acid**
   c. HF **hydrofluoric acid**
   d. HNO_3 **nitric acid**
   e. H_2SO_4 **sulfuric acid**

NOMENCLATURE: WRITING FORMULAS & NAMING CHEMICAL COMPOUNDS IONIC AND MOLECULAR (COVALENT) AND ACIDS

1. Write the chemical formulas for the following compounds:

   a. sodium acetate **$NaC_2H_3O_2$**
   b. ferric bicarbonate **$Fe(HCO_3)_3$**
   c. zinc sulfite **$ZnSO_3$**
   d. silver bicarbonate **$AgHCO_3$**
   e. potassium iodide **KI**
   f. barium bisulfate **BaHS**
   g. lead(IV) chlorite **$Pb(ClO_2)_4$**
   h. nitric acid **HNO_3**
   i. calcium sulfide **CaS**
   j. lead(II) nitrite **$Pb(NO_2)_2$**
   k. copper(I) bisulfate **$CuHSO_4$**
   l. potassium dichromate **$K_2Cr_2O_7$**
   m. sulfuric acid **H_2SO_4**
   n. boron monophosphide **BP**
   o. cobaltic chlorate **$Co(ClO_3)_3$**

2. Write the chemical names for the following compounds:

   a. SO_3 **sulfur trioxide**
   b. $Be(ClO_4)_2$ **beryllium perchlorate**
   c. $(NH_4)_2Cr_2O_7$ **ammonium dichromate**
   d. $Ba(BrO_3)_2$ **barium bromate**
   e. XeF_2 **xenon difluoride**
   f. Al_2S_3 **aluminum sulfide**
   g. Na_2HPO_4 **sodium biphosphate**
   h. $Mg_3(PO_4)_2$ **magnesium phosphate**
   i. $Al(OH)_3$ **aluminum hydroxide**
   j. $CuSO_3$ **copper (II) sulfite**
   k. Li_2HPO_4 **lithium biphosphate**
   l. $Ca(NO_3)_2$ **calcium nitrate**
   m. $Cr_2(SO_3)_3$ **chromic sulfite**
   n. $Ni(ClO_4)_2$ **nickel (II) perchlorate**
   o. HClO **hypochlorous acid**

UNIT 5: PRACTICE PROBLEMS

IONIC BONDING LEWIS DOT STRUCTURES

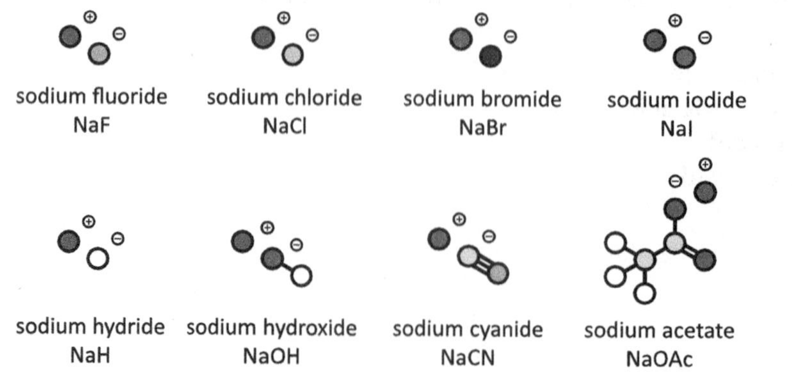

Ions are formed where electrons are transferred from the valence shell of one atom (usually a metal) to the valence shell of another atom (non-metal) so that both end up with noble gas configurations. Assume, in the first instance, that compounds between reactive metals and reactive non-metals will be ionic.

1. Draw diagrams (outer electrons only) to show the bonding in the following ionic compounds. Draw a before and after bond picture for each on another sheet of paper.

	Before	After
a. Lithium fluoride—LiF	Li· :F:	Li^{+} [:F:]$^{-}$
b. Magnesium sulphide—MgS	Mg: S:	Mg^{2+} [:S:]$^{2-}$
c. Calcium chloride—$CaCl_2$	Ca: 2.Cl:	Ca^{2+} 2[:Cl:]$^{-}$
d. Sodium oxide—Na_2O	2Na· :O:	2Na^{+} [:O:]$^{2-}$
e. Aluminium oxide—Al_2O_3	2Al: 3O:	2Al^{3+} 3[:O:]$^{2-}$
f. Magnesium nitride—Mg_3N_2	3Mg: 2N·	3Mg^{2+} 2[:N:]$^{3-}$

MOLECULAR BONDING LEWIS DOT STRUCTURE

Molecular bonding involves the sharing of electron pairs between two atoms. This occurs most often between non-metal atoms, but there are a number of compounds between metals and non-metals that are molecular. A single molecular bond involves one shared pair of electrons. In many compounds, atoms will share electrons to enable their valence shell to become like the nearest noble gas. This is normally 8 electrons (the Octet Rule), apart from hydrogen. There are exceptions (see next section).

Example: Making water

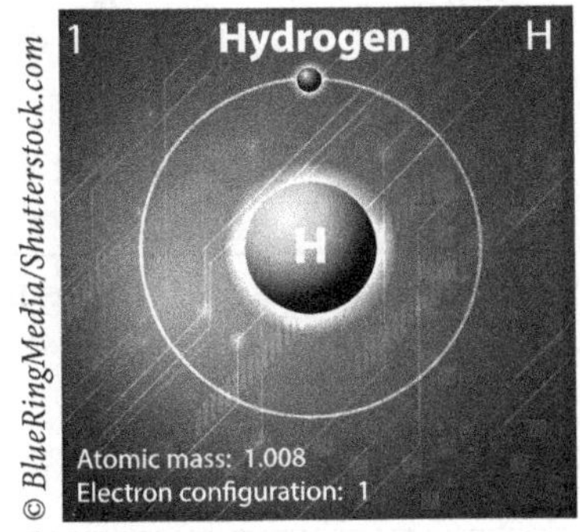

© BlueRingMedia/Shutterstock.com

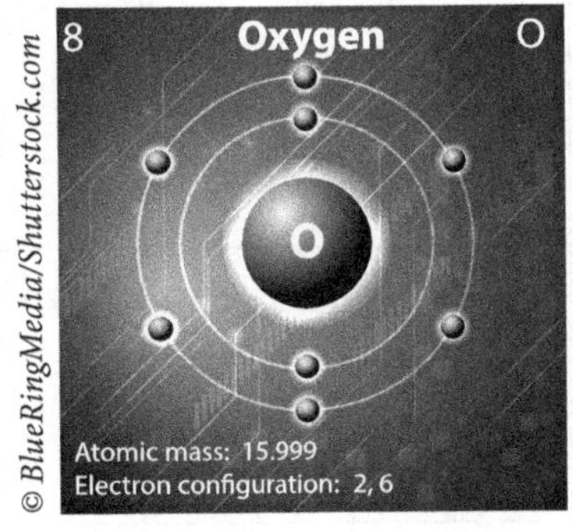

© BlueRingMedia/Shutterstock.com

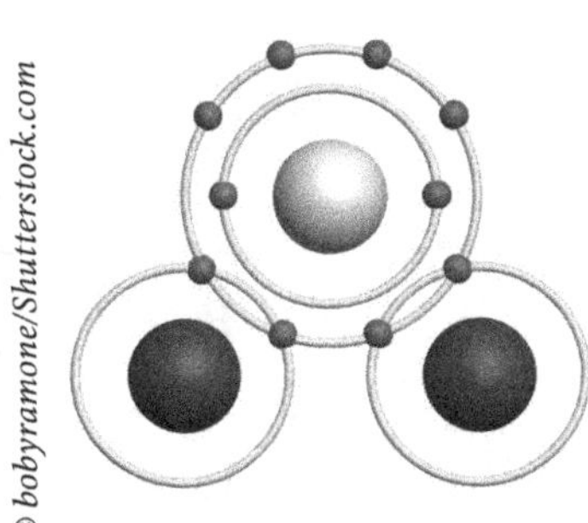
© bobyramone/Shutterstock.com

1. Draw diagrams (outer electrons only) to show the bonding in the following molecules. Draw a before and after bond picture for each on another sheet of paper. (Your drawings should be similar to the example above.)
   a. Hydrogen fluoride—HF **H–F̤̈:**
   b. Chlorine—Cl_2 **:C̤̈l–C̤̈l:**
   c. Oxygen—O_2 **Ö̤=Ö̤**
   d. Nitrogen—N_2 **:N≡N:**
   e. Cyanide—CN^- **[:C≡N:]⁻**

MOLECULAR GEOMETRY

Complete the Following Table:

Compound	Lewis Dot	Domain	Electron Pair	Molecular
CO_2	Ö̤=C=Ö̤	**AX_2**	**Linear**	**Linear**
BF_3	:F̤̈–B–F̤̈: with :F̤̈: bonded below B	**AX_3**	**Trigonal Planar**	**Trigonal Planar**
NO_2^-	[Ö̤=Ṅ̈–Ö̤:]⁻	**AX_2E**	**Trigonal Planar**	**Bent**
CH_4	H–C–H with H above and below C	**AX_4**	**Tetrahedral**	**Tetrahedral**

Compound	Lewis Dot	Domain	Electron Pair	Molecular
NH_3	H–N̈–H | H	AX_3E	**Tetrahedral**	**Trigonal Pyramidal**
H_2O	H–Ö–H	AX_2E_2	**Tetrahedral**	**Bent**
PCl_5	:Cl: Cl: :Cl–P–Cl: :Cl:	AX_5	**Trigonal Bipyramidal**	**Trigonal Bipyramidal**
SF_4	:F: :F–S–F: :F:	AX_4E	**Trigonal Bipyramidal**	**Seesaw**
ClF_3	:F–Cl–F: :F:	AX_3E_2	**Trigonal Bipyramidal**	**T-Shaped**
XeF_2	:F–Xe–F:	AX_2E_3	**Trigonal Bipyramidal**	**Linear**
SF_6	:F: F: :F–S–F: :F: :F:	AX_6	**Octahedral**	**Octahedral**
BrF_5	:F: F: :F–Br–F: :F:	AX_5E	**Octahedral**	**Square Pyramidal**
XeF_4	:F: :F–Xe–F: :F:	AX_4E_2	**Octahedral**	**Square Planar**

CHEMISTRY DISCOVER UNIT 8
LEWIS STRUCTURES AND MOLECULAR GEOMETRY

Compound	Lewis Dot	Domain	Electron Pair	Molecular
SO_2	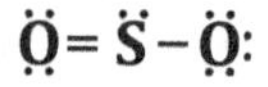 Ö=S̈–Ö:	AX_2E	**Trigonal planar**	**Bent**
BCl_3	:C̤̈l–B–C̤̈l: | :C̤̈l:	AX_3	**Trigonal planar**	**Trigonal planar**
NH_3	H–N̈–H | H	AX_3E	**Tetrahedral**	**Trigonal pyramidal**
PO_4^{3-}	[:Ö: | :Ö–P–Ö: | :Ö:]$^{3-}$	AX_4	**Tetrahedral**	**Tetrahedral**
NO_3^-	[:Ö–N=Ö: | :Ö:]$^{-}$	AX_3	**Trigonal Planar**	**Trigonal Planar**
SO_3	:Ö–S=Ö | :Ö:	AX_3	**Trigonal Planar**	**Trigonal Planar**
SO_3^{2-}	[:Ö–S̈–Ö: | :Ö:]$^{2-}$	AX_3E	**Tetrahedral**	**Trigonal Pyramidal**
PF_3	:F̈–P̈–F̈: | :F̈:	AX_3E	**Tetrahedral**	**Trigonal Pyramidal**
SF_6	:F̈: F̈: | / :F̈–S–F̈: / | :F̈: :F̈:	AX_6	**Octahedral**	**Octahedral**

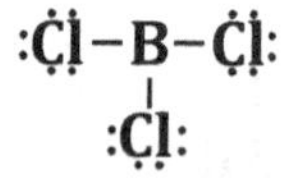

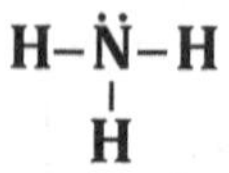

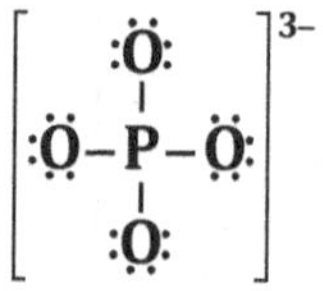

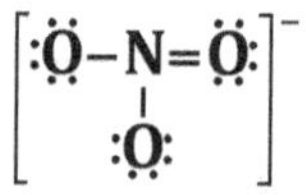

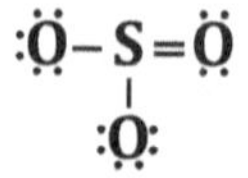

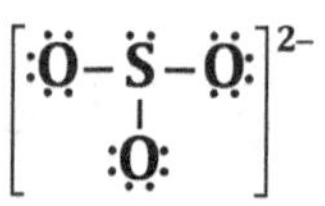

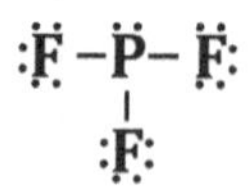

Compound	Lewis Dot	Domain	Electron Pair	Molecular
BrF_5	:F: F: :F – Br – F: :F:	**AX_5E**	**Octahedral**	**Square Pyramidal**
BrO_3^-	[:O – Br – O:]$^-$:O:	**AX_3E**	**Tetrahedral**	**Trigonal Pyramidal**
ICl_4	:Cl: :Cl – I – Cl: :Cl:	**AX_4E_2**	**Octahedral**	**Square Planar**
XeF_2	:F – Xe – F:	**AX_2E_3**	**Trigonal Bipyramidal**	**Linear**

ELECTRONEGATIVITY / POLARITY

Bonding Between	More Electronegative Element and Value	Less Electronegative Element and Value	Difference in Electronegativity	Bond Type
Sulfur and hydrogen	**S**	**H**	**0.4**	**Polar Molecular**
Sulfur and cesium	**S**	**Cs**	**1.7**	**50% Ionic** **50% Molecular**
Chlorine and bromine	**Cl**	**Br**	**0.2**	**Non-Polar Molecular**
Calcium and chlorine	**Cl**	**Ca**	**2.0**	**Ionic**
Oxygen and hydrogen	**O**	**H**	**1.4**	**Polar Molecular**
Nitrogen and hydrogen	**N**	**H**	**0.9**	**Polar Molecular**
Iodine and iodine	**Neither**	**Neither**	**0**	**Non-Polar Molecular**
Copper and sulfur	**S**	**Cu**	**0.7**	**Polar Molecular**

Bonding Between	More Electronegative Element and Value	Less Electronegative Element and Value	Difference in Electronegativity	Bond Type
Hydrogen and fluorine	**F**	**H**	**1.9**	**Ionic**
Carbon and oxygen	**O**	**C**	**1.0**	**Polar Molecular**

LEWIS DOT STRUCTURES AND MOLECULAR GEOMETRY

1. How many valence electrons does gallium have?
   a. 1
   b. 3
   c. 5
   d. 13

2. How many electrons does phosphorus have to gain in order to achieve a noble gas electron configuration?
   a. 2
   b. 3
   c. 5
   d. 4

   It has five valence electrons, and wants eight like argon.

3. How many valence electrons are transferred from the nitrogen atom to each potassium in the formation of the compound potassium nitride?
   a. 0
   b. 1
   c. 3
   d. 5

   Electrons are transferred from K to N, not the other way around.

4. Which of the following takes place in an ionic bond?
   a. Two atoms share two electrons
   b. Two atoms share electrons such that both follow the octet rule.
   c. Like-charged ions attract
   d. Oppositely-charged ions attract

5. What is the net charge of the ionic compound calcium fluoride?
   a. −2
   b. 0
   c. +2
   d. −1

   Ionic compounds have no overall charge, because the cations and anions cancel each other out.

6. How many electrons are shared between two atoms in a double molecular (covalent bond?)

   a. 8
   b. 6
   c. 4
   d. 2

   Because each bond contains two electrons, a double bond will contain 4 electrons.

7. How many unshared pairs of electrons are there in hydrogen iodide?

   a. 1
   b. 6
   c. 3
   d. none of these

   All three unshared electron pairs are on the iodine atom.

8. Which of the following elements occurs naturally as a diatomic molecule with a triple bond?

   a. oxygen
   b. hydrogen
   c. fluorine
   d. nitrogen

   The others are all naturally diatomic, but only N_2 has triple bonds.

9. Draw the 2-D Lewis structure below the molecular formula.

   - Determine both electron pair (EP) and molecular geometry.
   - From the overall molecular geometry and the presence and arrangement of polar bonds (if any), determine if a molecule is polar. (Polarity does not apply to polyatomic ions.)

   a. PF_3 **26e⁻**

   :F̤̈–P̈–F̤̈:
   |
   :F̤:

   EP geometry: **tetrahedral**
   Molecular geometry: **trigonal pyramidal**
   Is the molecule polar? **yes**

   b. PF_5 **40e⁻**

   :F̤̈:
   |
   :F̤̈–P–F̤̈:
   |
   :F̤: F̤:

   EP geometry: **trigonal bipyramidal**
   Molecular geometry: **trigonal bipyramidal**
   Is the molecule polar? **no**

c. SF_4

34e⁻

$:\ddot{F}:$ / $:\ddot{F}-\ddot{S}-\ddot{F}:$ / $:\ddot{F}:$

EP geometry: trigonal bipyramidal
Molecular geometry: see-saw
Is the molecule polar? yes

d. SF_6

48e⁻

$:\ddot{F}:\ddot{F}\cdot$ / $:\ddot{F}-\ddot{P}-\ddot{F}:$ / $\ddot{F}\ :\ddot{F}:$

EP geometry: octahedral
Molecular geometry: octahedral
Is the molecule polar? no

e. CH_3^+

6e⁻

$[H-C-H]^+$ (uncommon for carbon)

EP geometry: trigonal planar
Molecular geometry: trigonal planar
Is the molecule polar? n/a

f. ClO^-

14e⁻

$[:\ddot{Cl}-\ddot{O}:]^-$

EP geometry: linear
Molecular geometry: linear
Is the molecule polar? n/a

g. ClO_2^-

20e⁻

$[:\ddot{O}-:\ddot{Cl}-\ddot{O}:]^-$

EP geometry: tetrahedral
Molecular geometry: bent
Is the molecule polar? n/a

h. ClO_3^-

26e⁻

$[:\ddot{O}-\dot{Cl}-\ddot{O}: \; :\ddot{O}:]^-$

EP geometry: tetrahedral
Molecular geometry: trigonal pyramidal
Is the molecule polar? n/a

i. ClO_4^-

32e⁻

$[:\ddot{O}: \; :\ddot{O}-\dot{Cl}-\ddot{O}: \; :\ddot{O}:]^-$

EP geometry: tetrahedral
Molecular geometry: tetrahedral
Is the molecule polar? n/a

j. KrF_2

$22e^-$

:F–Kr–F:

EP geometry: **trigonal bipyramidal**
Molecular geometry: **linear**
Is the molecule polar? **no**

k. XeF_4

$36e^-$

:F:
|
:F–Xe–F:
|
:F:

EP geometry: **octahedral**
Molecular geometry: **square planar**
Is the molecule polar? **no**

l. XeO_3

$26e^-$

:O–Xe–O:
|
:O:

EP geometry: **tetrahedral**
Molecular geometry: **trigonal pyramidal**
Is the molecule polar? **yes**

m. XeO_2F_2

$34e^-$

:F:
|
:O–Xe–O:
|
:F:

EP geometry: **trigonal bipyramidal**
Molecular geometry: **see-saw**
Is the molecule polar? **yes**

n. CS_2

$16e^-$

S=C=S

EP geometry: **linear**
Molecular geometry: **linear**
Is the molecule polar? **no**

o. NO_3^-

$24e^-$

:O–N–O:
||
:O:

EP geometry: **trigonal planar**
Molecular geometry: **trigonal planar**
Is the molecule polar? **n/a**

p. CO_3^{2-}

$24e^-$

[:O–C–O: ; || ; :O:]$^{2-}$

EP geometry: **trigonal planar**
Molecular geometry: **trigonal planar**
Is the molecule polar? **n/a**

q. BH_3

6e⁻

```
H–B–H
  |
  H
```

EP geometry: **trigonal planar**
Molecular geometry: **trigonal planar**
Is the molecule polar? **no**

r. XeF_3^+

28e⁻

```
:F–Xe–F:
    |
   :F:
```

EP geometry: **trigonal bipyramidal**
Molecular geometry: **T-shaped**
Is the molecule polar? **n/a**

s. BrF_3

28e⁻

```
:F–Br–F:
    |
   :F:
```

EP geometry: **trigonal bipyramidal**
Molecular geometry: **T-shaped**
Is the molecule polar? **yes**

t. IF_5

42e⁻

```
    :F:
     |
:F – I – F:
     | \
    :F: F:
```

EP geometry: **octahedral**
Molecular geometry: **square planar**
Is the molecule polar? **yes**

u. HCN

10e⁻

```
H–C≡N:
```

EP geometry: **linear**
Molecular geometry: **linear**
Is the molecule polar? **yes**

v. NH_3Cl^+

14e⁻

```
[    H     ]+
[    |     ]
[ H–N–Cl:  ]
[    |     ]
[    H     ]
```

EP geometry: **tetrahedral**
Molecular geometry: **tetrahedral**
Is the molecule polar? **n/a**

w. I_3^-

22e⁻

```
[:I–I–I:]⁻
```

EP geometry: **trigonal bipyramidal**
Molecular geometry: **linear**
Is the molecule polar? **n/a**

x. N_3^- 16e⁻

$[\dot{N}=N=\dot{N}]^-$

EP geometry: linear
Molecular geometry: linear
Is the molecule polar? n/a

y. $BHCl_2$ 18e⁻

H
|
:Cl–B–Cl:

EP geometry: trigonal planar
Molecular geometry: trigonal planar
Is the molecule polar? yes

UNIT 6: PRACTICE PROBLEMS

ATOMIC MASS / MOLECULAR MASS / FORMULA MASS / MOLAR MASS

1. Determine the atomic mass, formula mass, or molecular mass for each of the following (indicate if it is either a(n) atomic mass, formula mass, or molecular mass):
   a. $BaSO_4$ 233.39 amu (Formula Mass)

   b. N_2H_4 32.05 amu (Molecular Mass)

   c. aluminum oxalate 318.02 amu (Formula Mass)

   d. acetic acid 60.05 amu (Molecular Mass)

   e. arsenic 74.92 amu (Atomic Mass)

MOLAR MASS

1. Determine the molar mass for each of the following:

   a. ZnI_2 319.22 g/mole

   b. NO_2 46.01 g/mole

   c. Iron(II) nitrate 179.86 g/mole

MOLE CONVERSIONS

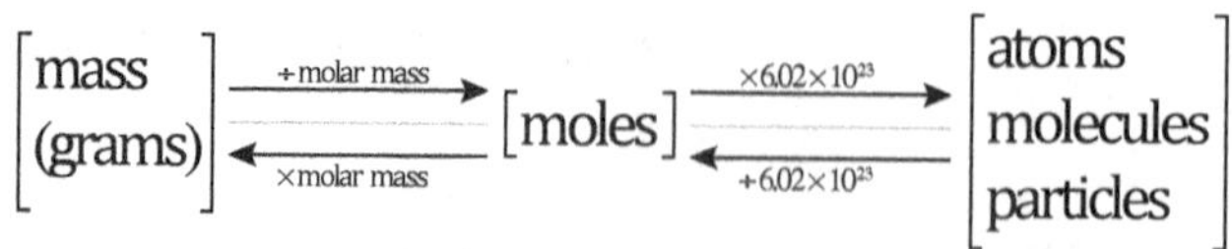

1. Convert mass (grams) to moles.
   Example: Determine the amount of moles in 40.0 grams Mg.
   Set up:

Try:

4.50 grams Fe = **0.0806 mol Fe**	0.632 grams C_6H_6 = **8.10 × 10^{-3} mol C_6H_6**
1.6 × 10^{12} grams Rb = **1.9 × 10^{10} mol Rb**	52.670 grams barium chloride = **0.25286 mol $BaCl_2$**
32.12 grams ammonium sulfide = **0.4717 mol $(NH_4)_2S$**	15.6 grams $(NH_4)_2SO_4$ = **0.118 mol $(NH_4)_2SO_4$**

2. Convert moles to mass (grams).
   Example: Determine the mass of 6.43 moles C_2H_4
   Set up:

Try:

15.67 moles Cr = **814.8 g Cr**	0.00032 moles CO_2 = **0.014 g CO_2**
4.56 × 10^{12} moles S = **1.46 × 10^{14} g S**	35,067 moles beryllium nitrate = **4,663,900 g $Be(NO_3)_2$**
32,000 moles beryllium nitride = **1,800,000 g Be_3N_2**	2.00 moles Ne = **40.2 g Ne**

3. Convert moles to atoms, molecules or particles.
   Example: Determine the number of particles in 0.677 moles $NaHCO_3$
   Set up:

Try:

5.6 × 10^{3} moles Ba = **3.4 × 10^{27} atoms Ba**	7.64 × 10^{-1} moles dihydrogen monoxide = **4.60 × 10^{23} molecules H_2O**
0.4428 moles $CaCO_3$ = **2.666 × 10^{23} particles $CaCO_3$**	4,560 moles cobalt (II) dichromate = **2.75 × 10^{27} particles $CoCr_2O_7$**
8.21 moles XeF_4 = **4.94 × 10^{24} molecules XeF_4**	2.00 moles NO = **1.20 × 10^{24} molecules NO**

4. Convert atoms, molecules, or particles to moles.
   Example: Determine the mole quantity in of 5.6×10^{24} atoms Ag.
   Set up:

Try:

4.98×10^{23} particles $MoBr_2$ = **0.827 mol $MoBr_2$**	4.88×10^{21} atoms Ra = **8.11×10^{-3} mol Ra**
2.3333×10^{24} atoms Si = **3.8759 mol Si**	5.1×10^{24} molecules H_2O_2 = **8.5 mol H_2O_2**
5.882×10^{26} molecules P_5S_7 = **977.1 mol P_5S_7**	7×10^{23} particles MgO = **1 mol MgO**

5. Convert mass (grams) to atoms, molecules, or particles.
   Examples: How many molecules of alcohol are there in a "standard" can of beer if there are 21.3 g of C_2H_6O?
   Set up:

Try:

35.899 grams Au = **1.0970×10^{23} atoms Au**	0.333 grams NH_4OH = **5.73×10^{21} particles NH_4OH**
54.5 grams Na_2Se = **2.62×10^{23} particles Na_2Se**	65.6 grams At_2 = **9.40×10^{22} molecules At_2**
5.650 grams $SrSO_4$ = **1.852×10^{22} particles $SrSO_4$**	4.500 grams calcium bromide = **1.355×10^{22} particles $CaBr_2$**

6. Convert atoms, molecules, or particles to grams.
   Example: What is the mass of 5.888×10^{25} molecules C_3H_8?
   Set up:

Try:

4.56×10^{24} atoms Ag = **817 g Ag**	5.6000×10^{20} molecules $C_6H_{12}O_6$ = **0.16744 g $C_6H_{12}O_6$**
1.230×10^{25} particles BeO = **510.8 g BeO**	7.2×10^{14} particles $Sn(CrO_4)_2$ = **4.2×10^{-7} g $Sn(CrO_4)_2$**
3.20×10^{32} atoms Hg = **1.07×10^{11} g Hg**	1.54×10^{22} atoms Po = **5.35 g Po**

PERCENT COMPOSITION

1. Determine the percentage composition of the elements in each of the following compounds:
   a. barium arsenate ____ **59.7% Ba, 21.7% As, 18.6% O**

   b. CH_3COOH ____ **40.0% C, 53.3% O, 6.67% H**

   c. Ammonium carbonate ____ **29.2% N, 8.33% H, 12.5% C, 50.0% O**

EMPIRICAL AND MOLECULAR FORMULAS

1. Determine the empirical formula for each of the following:
   a. CH_3COOH ____ **CH_2O**

   b. $C_6H_{12}O_6$ ____ **CH_2O**

   c. $N_2O_{10}F_5$ ____ **$N_2O_{10}F_5$**

2. Determine the empirical formula for each of the following:

   30.44% N, 69.55% O, molecular mass = 92 amu = **NO_2 / N_2O_4**

   9.93% C, 58.64% Cl, 31.43% F, molecular mass = 121 amu = **CCl_2F_2 / CCl_2F_2**

   40.00% C, 6.71% H, 53.29% O, molecular mass = 60 amu = **CH_2O / $C_2H_4O_2$**

MOLE CALCULATIONS

Tin

87.54 grams Sn convert to moles

0.7374 mol Sn

Table sugar

0.143 mole $C_{12}H_{22}O_{11}$ convert to grams

48.9 g $C_{12}H_{22}O_{11}$

Mothballs

1.35×10^{23} molecules $C_{10}H_8$ (naphthalene) convert to moles

0.224 mol $C_{10}H_8$

Ice-melting salts

0.156 mole $CaCl_2$ convert to particles

9.39×10^{22} particles $CaCl_2$

Baking soda

23.0 grams $NaHCO_3$ convert to particles

1.65×10^{23} particles $NaHCO_3$

Water

2.5×10^{24} molecules of water convert to grams

75 grams water

REVIEW CALCULATIONS

1. Calculate the amount of molecules in 35.5 grams citric acid (acid found in citrus fruits) $C_6H_8O_7$.
   1.11×10^{23} molecules

2. What is the mass of 4.59×10^{29} particles of MgO?
   3.07×10^{7} grams

3. Is sodium nitrate a molecule or a particle? Explain.
   Particle—ionic compound

4. Calculate the formula mass of ammonium phosphate.
   149.0858 amu

5. Calculate the molecular mass of C_2H_4OH.
   45.0605 amu

6. Which of the following compound is the term “molecular mass” better suited than the term “formula mass”?
   a. F_2
   b. CaO
   c. K_3N
   d. NaOH
   e. RbOH

7. Calculate the number of moles in 456.8 grams of $Al(OH)_3$.
   5.856 mol

8. Calculate the molar mass of N_2S
   60.078 g/mol

9. What is the mass of 0.0490 mole of titanium(IV) oxide?
   3.91 g

10. How many atoms are in 18.8 mol Rb?
    1.13×10^{25} atoms

11. Which of the following statements is incorrect?
    a. Molecular mass is expressed in amu.
    b. Formula mass is expressed in amu.
    c. Atomic mass is expressed in grams/atom.
    d. Molar mass is expressed in grams/mole.

12. Calculate the mass of 6.78×10^{31} particles $Mg_3(PO_4)_2$.
    2.96×10^{10} grams

13. How many molecules are in 20.35 mol NH_3?
    1.225×10^{25} molecules

14. How many atoms are in 44.3 moles of barium?
    2.67×10^{25} atoms

15. Calculate the number of atoms in 566 grams of carbon.
    2.84×10^{25} atoms

16. How many grams are in 54.87 mol of CaH_2?
    2310. grams

17. How many moles are in 6.27×10^{24} particles of $NaNO_3$?
    10.4 mol

18. Calculate the mass of 7.99×10^{25} particles Fe_2O_3.

    21200 grams

19. Find the percent composition of ammonium nitrate.

    35% N, 5% H, 60% O

20. NutraSweet is 57.14% C, 6.16% H, 9.52% N, and 27.18% O. Calculate the empirical formula of NutraSweet and find the molecular formula. (The molar mass of NutraSweet is 294.30 g/mol.)

    Empirical: $C_{14}H_{18}N_2O_5$
    Molecular: $C_{14}H_{18}N_2O_5$

ATOMIC, FORMULA, AND MOLECULAR FORMULAS & MASSES

1. Determine the formula or molecular masses for the following compounds:
    a. $Mg(OH)_2$ = **58.3 amu**
    b. calcium hydroxide = **74.1 amu**
    c. acetic acid = **60.0 amu**

2. The controversial artificial sweetener saccharin has the molecular formula $C_3H_5O_3NS$. What is its molecular mass?
    a. 123.12 amu
    b. 119.88 amu
    c. 135.14 amu
    d. 103.15 amu
    e. 77.78 amu

3. Cisplatin, an anticancer drug, has the molecular formula $Pt(NH_3)_2Cl_2$. What is the molecular mass of cisplatin?
    a. 323.3 amu
    b. 300.1 amu
    c. 332.6 amu
    d. 321.2 amu
    e. 340.4 amu

4. Nitroglycerin is $C_3H_5N_3O_9$. What is the molecular mass of nitroglycerin?
    a. 240.22 amu
    b. 286.44 amu
    c. 227.10 amu
    d. 270.42 amu
    e. 256.20 amu

5. The formula mass of calcium hydroxide, $Ca(OH)_2$ is:
    a. 128 amu
    b. 74 amu
    c. 97 amu
    d. 57 amu

6. The formula mass of magnesium hydroxide, $Mg(OH)_2$ is:
   a. 42.33 amu
   b. 58.33 amu
   c. 41.32 amu
   d. 5 amu

7. The total number of OXYGEN atoms in the formula of aluminum dichromate, $Al_2(Cr_2O_7)_3$ is:
   a. 10
   b. 7
   c. 29
   d. 21

8. How many atoms are in 18 molecules of glucose, $C_6H_{12}O_6$?
   a. 24
   b. 432
   c. 3240
   d. 1.08×10^{25}
   e. 2.60×10^{26}

9. Which of the following samples contains the smallest number of molecules?
   a. 1 g phosphorus, P_4
   b. 1 g chlorine, Cl_2
   c. 1 g nitrogen, N_2
   d. 1 g arsenic, As_4
   e. 1 g sulfur, S_8

MOLAR MASS

1. The molar mass of sodium chloride, NaCl is:
   a. 58.44 g/mol
   b. 69.71 g/mol
   c. 2 g/mol
   d. 6.022×10^{23} g/mol

THE MOLE CALCULATIONS

1. What is the mass in grams of 10. moles of ammonia, NH_3?
   a. 170 grams
   b. 27.0 grams
   c. 1.70 grams
   d. 0.59 grams

2. $BaSO_4$ is given as a thick slurry before X-rays are taken of the intestinal tract. How many grams are in a 0.568 mole sample of $BaSO_4$?
   a. 62.4
   b. 103
   c. 56.8
   d. 77.8
   e. 133

3. Which of the following correctly describes the mole?
   a. One mole is 6.022×10^{23} atoms of any element.
   b. One mole is the number of atoms in exactly 12.0 g of ^{12}C.
   c. One mole of any chemical compound is one mole of its chemical formula unit.
   d. All of the above are correct.

4. About how many atoms of helium would be found in 2 grams of helium?
   a. 6×10^{23}
   b. 4
   c. 3×10^{23}
   d. 2

5. What is the mass of 4 moles of hydrogen molecules (H_2)?
   a. 4 grams
   b. 3 grams
   c. 8 grams
   d. 1 gram

6. What is the mass in grams of 3.00 moles of water molecules, H_2O?
   a. 54.0 grams
   b. 21.0 grams
   c. 6.01 grams
   d. 0.166 grams

7. How many moles of water molecules, H_2O, are present in a 42.0 gram sample of water?
   a. 2.33 moles
   b. 0.429 moles
   c. 23.98 moles
   d. 757 moles

8. What is the mass of 1.004×10^{23} molecules of barium iodide?
   a. 44.05 g
   b. 44.12 g
   c. 65.20 g
   d. 0.167 g

9. How many moles are in 32.0 grams of CH_4?
   a. 32.0 moles
   b. 16.0 moles
   c. 1.00 mole
   d. 2.00 moles

10. How many moles of methane molecules, CH_4, are in 80 grams of methane?
    a. 1284 moles
    b. 6×10^{80} moles
    c. 0.2 moles
    d. 5 moles

11. The amount of substance having 6.022×10^{23} of any kind of chemical unit is called a(n):
    a. mole
    b. mass number
    c. atomic weight
    d. formula

12. Determine the mass of one mole of CO_2.
44.0 g CO_2

13. Determine the number of atoms in one molecule of Fe_2O_3.
2 atoms Fe, 3 atoms O = 5 atoms

14. Determine the number of atoms in one mole of $C_6H_{12}O_6$.
6.02×10^{23} molecules $C_6H_{12}O_6$ = 3.61×10^{24} atoms C, 7.22×10^{24} atoms H, 3.61×10^{24} atoms O

15. Determine the mass of 5.240 moles of gold.
1032 g Au

16. Determine the number of moles of nitrogen gas in 85.4 g of nitrogen gas.
3.05 mol N_2

17. Determine the mass, in grams, of one atom of silver.
1.79×10^{-22} g Ag

18. Determine the mass, in grams, of one molecule of carbon dioxide.
7.31×10^{-23} g CO_2

19. How many molecules are there in 10.0 g of sodium chloride?
    a. 58.4×10^{23}
    b. 5.84×10^{24}
    c. 6.02×10^{24}
    d. 1.03×10^{23}

20. In 0.250 moles of ethylene glycol (antifreeze), $HOCH_2CH_2OH$, there are:
    a. 1.51×10^{23} atoms
    b. 1.51×10^{24} molecules
    c. 1.51×10^{24} atoms
    d. 6.02×10^{24} atoms
    e. 3.01×10^{24} molecules

21. Which of the following does not describe 56.0 g of butene, C_4H_8?
    a. One mole of butene
    b. The amount of butene that contains 8.0 g of hydrogen
    c. The amount of butene that contains $8 \times 6.02 \times 10^{23}$ hydrogen atoms
    d. The amount of butene that contains 48.0 g of carbon
    e. $56.0 \times 6.02 \times 10^{23}$ molecules of butene

22. Sodium cyclamate, $C_6H_{11}NHSO_3Na$, is used as an artificial sweetener in South Africa. If $C_6H_{11}NHSO_3Na$ has a molar mass of 201.2 g/mol, how many moles of sodium cyclamate are contained in a 25.6 g sample?
    a. 0.127 mol
    b. 0.193 mol
    c. 0.245 mol
    d. 7.90 mol
    e. 5180 mol

23. How many moles of nitrogen gas (N_2 molecules) are present in 48.0 grams of nitrogen?
    a. 0.58 mol
    b. 0.86 mol
    c. 1.71 mol
    d. 2.00 mol
    e. 3.42 moles

24. Which one of the following has the lightest mass?
    a. An HF molecule
    b. 20.0 g of HF
    c. 10.0 mol of H_2
    d. 1 mol of F_2
    e. 1 mol of H_2O

25. One mole is ______.
    a. the amount of molecules in any substance
    b. the amount of particles in any substance
    c. the amount of atoms in any substance
    d. the amount of ions in any substance
    e. just a number

PERCENT COMPOSITION

1. Determine the percent composition of ammonium sulfate, $(NH_4)_2SO_4$.
   21% N, 6.1% H, 24.3% S, 48.4% O

2. Determine the percent composition of urea, N_2H_4CO.
   46.7% N, 6.7% H, 20.0% C, 26.7% O

3. Each of the compounds listed in questions 1 & 2 contains nitrogen. They are used as fertilizers. For each of the compounds, which one has the highest percentage of nitrogen?
   N_2H_4CO

4. Calculate the percentage composition of $Ca(ClO_3)_2$?
   a. 19.4% Ca, 34.3% Cl, 46.4% O
   b. 32.4% Ca, 28.7% Cl, 38.8% O
   c. 49.0% Ca, 21.8% Cl, 29.3% O
   d. 32.4% Ca, 67.6% ClO_3
   e. 19.4% Ca, 51.0% ClO_3

EMPIRICAL & MOLECULAR FORMULAS

1. Write the empirical formula for each of the following compounds.
   a. $C_{12}H_{22}O_{11}$, sugar
      $C_{12}H_{22}O_{11}$

   b. $C_2H_6O_2$, ethylene glycol (antifreeze)
      CH_3O

2. From the following empirical formulas and the formula masses for each compound, determine their molecular formulas.
   a. CH_3; formula mass = 30.0 amu
      C_2H_6

   b. CH_2; formula mass = 84.0 amu
      C_6H_{12}

   c. $C_3H_4O_3$ (Vitamin C); formula mass = 176 amu
      $C_6H_8O_6$

3. Determine the empirical formula for a compound that contains 18.6 grams of phosphorus and 14.0 grams of nitrogen.
   P_3N_5

4. Determine the empirical formula for a compound that contains 35.6% of phosphorus and 64.4% of sulfur.
   P_4S_7

5. A compound with a molecular mass of 98.0 g/mole was determined to be 24.49% carbon, 4.08% hydrogen, and 72.43% chlorine.
   a. Determine the empirical formula of the compound.
      CH_2Cl

   b. Determine the molecular formula of the compound.
      $C_2H_4Cl_2$

6. What is the empirical formula for N_8O_4?
   a. NO_2
   b. NO
   c. N_2O
   d. ON_4
   e. N_2O_2

UNIT 7: PRACTICE PROBLEMS

Give an Example for Each of the Following:

Combination (CA) Decomposition (D) Combustion (CU)

Single Replacement (SR) Double Replacement (DR)

Chemical Reactions	Classify
1. Magnesium burning Write the chemical reaction. Identify the reactants and products. Identify the phase labels. Balance the equation **$2Mg(s) + O_2(g) \rightarrow 2MgO(s)$**	**CA**
2. copper + silver nitrate → silver + copper (II) nitrate Write the chemical reaction. Identify the reactants and products. Identify the phase labels. Balance the equation **$Cu(s) + 2AgNO_3(aq) \rightarrow 2Ag(s) + Cu(NO_3)_2(aq)$**	**SR**
3. Explain at the particulate (in molecule ratio) and the macroscopic (in mole ratio), what the coefficients represent in the combination of hydrogen and oxygen to make water. **2 molecules H_2 + 1 molecule O_2 → 2 molecules H_2O** **2 moles H_2 + 1 mole O_2 → 2 moles H_2O**	**CA**
4. $Mg(s) + ZnCl_2(aq) \rightarrow MgCl_2(aq) + Zn(s)$	**SR**
5. _2_ $AgNO_3(aq)$ + CaCl2(aq) → _2_ $AgCl(s) + Ca(NO_3)_2(aq)$	**DR**
6. _2_ $C_2H_6(g)$ + _7_ $O_2(g)$ → _4_ $CO_2(g)$ + _6_ $H_2O(g)$	**CU**
7. $Na_2O(s) + H_2O(l)$ → _2_ $NaOH(aq)$	**CA**
8. _2_ $KClO_3(s)$ → _2_ $KCl(s)$ + _3_ $O_2(g)$	**D**
9. _2_ $Al(s)$ + _6_ $HCl(aq)$ → _2_ $AlCl_3(aq)$ + _3_ $H_2(g)$	**SR**
10. $C_3H_6O(g)$ + _4_ $O_2(g)$ → _3_ $CO_2(g)$ + _3_ $H_2O(g)$	**CU**

Reaction	Classify
11. __4__ $Fe(s)$ + __3__ $O_2(g) \rightarrow$ __2__ $Fe_2O_3(s)$	CA
12. $Cl_2(aq)$ + __2__ $KBr(aq) \rightarrow Br_2(aq)$ + __2__ $KCl(aq)$	SR
13. __3__ $Ca(NO_3)_2(aq)$ + __2__ $K_3PO_4(aq) \rightarrow Ca_3(PO_4)_2(s)$ + __6__ $KNO_3(aq)$	DR
14. $Ca(HCO_3)_2(aq) \rightarrow CaCO_3(s) + H_2O(l) + CO_2(g)$	D
15. __3__ $NaOH(aq) + H_3PO_4(aq) \rightarrow Na_3PO_4(s)$ + __3__ $H_2O(l)$	DR
16. $BaCl_2(aq) + Na_2SO_4(aq) \rightarrow BaSO_4(s)$ + __2__ $NaCl(aq)$	DR
17. $Fe(s) + CuCl_2(aq) \rightarrow Cu(s) + FeCl_2(aq)$	SR
18. __2__ $C_6H_6(l)$ + __15__ $O_2(g) \rightarrow$ __12__ $CO_2(g)$ + __6__ $H_2O(l)$	CU
19. $BaCl_2(s) + H_2O(l) \rightarrow BaCl_2 \bullet H_2O(s)$	CA
20. $Pb(NO_3)_2(aq) + K_2CrO_4(aq) \rightarrow PbCrO_4(s)$ + __2__ $KNO_3(aq)$	DR
21. $Mg(s) + H_2SO_4(aq) \rightarrow MgSO_4(aq) + H_2(g)$	SR
22. $CaO(s) + H_2O(l) \rightarrow Ca(OH)_2(s)$	CA
23. __2__ $C_2H_4O(l)$ + __5__ $O_2(g) \rightarrow$ __4__ $CO_2(g)$ + __4__ $H_2O(l)$	CU
24. __2__ $HNO_3(aq) + Ba(OH)_2(aq) \rightarrow Ba(NO_3)_2(aq)$ + __2__ $H_2O(l)$	DR

STOICHIOMETRY

THE DECOMPOSITION OF HYDROGEN PEROXIDE (H_2O_2)

7. What do you think the bubbles or gas are?
   Oxygen and water are released in the decomposition of hydrogen peroxide.

8. What do you think causes the bubbling when hydrogen peroxide is placed on an open wound?
   Catalyze enzymes are catalysts found in cells that initiate the decomposition of the hydrogen peroxide.

	Reactants	→	Products		
equation	$2\ H_2O_2$	→	$2\ H_2O$	+	O_2
mole ratio	**2 moles H_2O_2**	→	**2 moles H_2O**	+	**1 mole O_2**
molar mass	**H_2O_2 = 34.02 g/mole**		**H_2O = 18.02 g/mole**		**O_2 = 32.00 g/mole**
gram ratio	**68.04 grams H_2O_2**	→	**36.04 grams H_2O**	+	**32.00 grams O_2**

INVESTIGATION OF THE SUGAR MOLECULE

1. The molecule below is a sucrose or sugar molecule. Use lines to connect each water molecule to prove that there are 11 water molecules in sucrose.

Sucrose (saccharose)

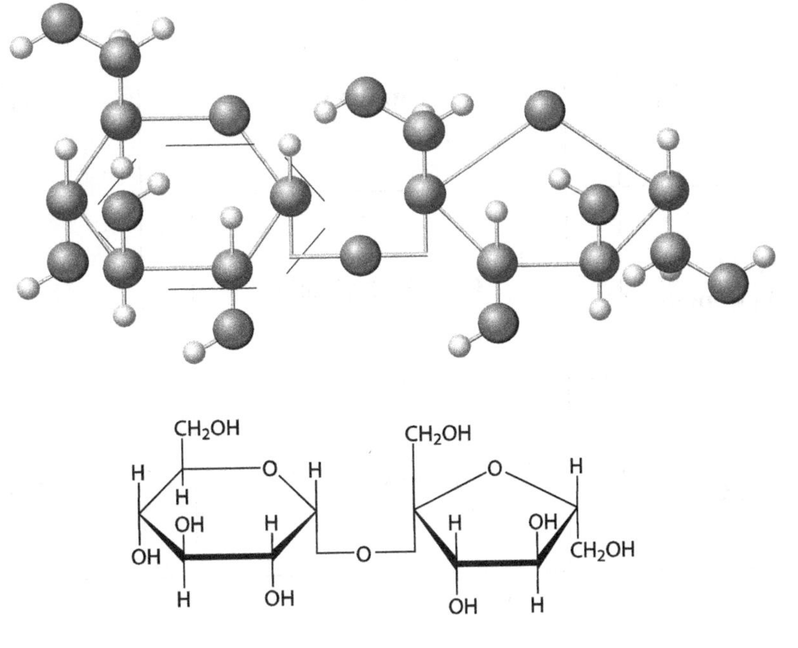

2. The molecule below is a simple carbohydrate called glucose or blood sugar. What is the formula for glucose?

   $C_6H_{12}O_6$

3. Place circles around hydrogen and oxygen atoms, then use lines to connect 2 hydrogen and 1 oxygen atoms to prove there are 6 water molecules.

4. Prove that sucrose is made up of two glucose-like molecules minus one H_2O molecule.

α-D-Glucose (cyclic)

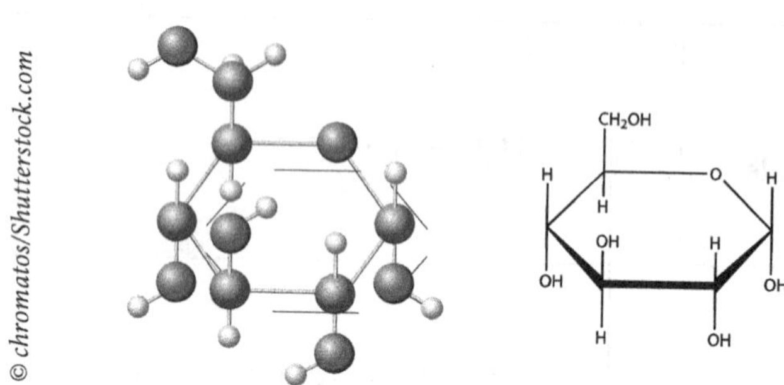

Data Table: Decomposition of sucrose (demonstrated)

	Reactants	→	Products		
equation	$C_{12}H_{22}O_{11}$	→	**12 C**	+	**11 H_2O**
mole ratio	**1 mole $C_{12}H_{22}O_{11}$**	→	**12 moles C**	+	**11 moles H_2O**
molar mass	**$C_{12}H_{22}O_{11}$ = 342.34 g/mole**		**C = 12.01 g/mole**		**H_2O = 18.02 g/mole**
gram ratio	**342.34 grams $C_{12}H_{22}O_{11}$**	→	**144.12 grams C**	+	**198. 22 grams H_2O**

STOICHIOMETRY

A + B → C

MOLE A → MOLE C

1. Combustion of Propane (C_3H_8).

   $C_3H_8(g) + 5O_2(g) \rightarrow 3CO_2(g) + 4H_2O(g)$

   Calculate the number of moles of oxygen reacting if 4.56 moles of CO_2 is produced.

   7.60 mol O_2

2. How many moles of oxygen are consumed when 32.4 moles of carbon dioxide is formed during the combustion of C_7H_{16}?

   $C_7H_{16} + 11O_2 \rightarrow 7CO_2 + 8H_2O$

   50.9 mol O_2

3. How many moles of $C_6H_{12}O_6$ are formed when 0.250 mole of CO_2 is consumed in the reaction

   $6CO_2 + 6H_2O \rightarrow C_6H_{12}O_6 + 6O_2$

   0.0417 mol $C_6H_{12}O_6$

MOLE B → GRAM C

1. Combustion of Propane (C_3H_8).

   $C_3H_8(g) + 5O_2(g) \rightarrow 3CO_2(g) + 4H_2O(g)$

   What is the mass of propane reacting if 765 moles of water is produced?

   8430 g C_3H_8

2. In the combustion of natural gas according to the equation

   $CH_4 + 2O_2 \rightarrow CO_2 + 2H_2O$

   how many grams of water are formed during the combustion of 0.264 mole of CH_4?

   9.50 g H_2O

3. In the equation: $I_2 + 7F_2 \rightarrow 2IF_7$

   3.28 moles of fluorine will yield ______ grams of IF_7.

   244 g IF_7

GRAM C → MOLE A

1. Combustion of Propane (C_3H_8).

   $C_3H_8(g) + 5O_2(g) \rightarrow 3CO_2(g) + 4H_2O(g)$

   How many moles of carbon dioxide are produced from 56.8 grams of oxygen reacting?

   1.07 mol O_2

2. In the complete combustion of $C_3H_8O_3$, how many moles of carbon dioxide are produced when 23.0 g of $C_3H_8O_3$ burns?

   $2C_3H_8O_3 + 7O_2 \rightarrow 6CO_2 + 8H_2O$

   0.750 mol CO_2

3. How many moles of bromine will react with 6.50 grams of C_2H_2 in the reaction

   $C_2H_2 + 2Br_2 \rightarrow C_2H_2Br_4$

   0.500 mol Br_2

GRAM C → GRAM A

1. Combustion of Propane (C_3H_8).

   $C_3H_8(g) + 5O_2(g) \rightarrow 3CO_2(g) + 4H_2O(g)$

   What is the mass of propane reacting to produce 3.43 grams of water?

   2.10 g C_3H_8

2. According to the following equation:

   $4NH_3 + 5O_2 \rightarrow 4NO + 6H_2O$

   How many grams of NO will form when 40. grams of oxygen are used?

   30. g NO

3. In the equation:

   $N_2 + 2O_2 \rightarrow 2NO_2$

   64 grams of oxygen will yield ____ grams of nitrogen dioxide.

   92 g NO_2

STOCHIOMETRY PROBLEMS

1. How many moles of chlorine gas (Cl_2) would react with 5.0 moles of sodium (Na) according to the following chemical equation? (Balance equation.)

   ____ Na + ____ Cl_2 → ____ NaCl

   2.5 mol Cl_2

2. If you start with 10.0 grams of aluminum hydroxide, how many grams of water will be produced?

   $Al(OH)_3 + 3HBr \rightarrow AlBr_3 + 3H_2O$

   6.92 grams

3. If you start with 4.500 moles of ethylene (C_2H_4), how many grams of carbon dioxide will be produced?

   $C_2H_4 + 3O_2 \rightarrow 2CO_2 + 2H_2O$

   396 grams

4. If you start with 5.5 grams of NaF, how many moles of magnesium fluoride will be produced?

   $Mg + 2NaF \rightarrow MgF_2 + 2Na$

   0.065 moles

5. If you start with 20.0 grams of hydrochloric acid, how many grams of sulfuric acid will be produced?

   $2HCl + Na_2SO_4 \rightarrow 2NaCl + H_2SO_4$

   26.9 grams

6. $C_3H_8 + 5O_2 \rightarrow 3CO_2 + 4H_2O$
   a. If I start with 5.0 grams of C_3H_8, what is my theoretical yield of water?
      8.2 grams

   b. Calculate the percentage yield if 6.24 grams of water are actually obtained by experiment (with starting with 5.0 grams C_3H_8).
      76.1%

7. My theoretical yield of beryllium chloride was 10.7 grams. If my actual yield was 4.5 grams, what was my percentage yield?

   $Be + 2HCl \rightarrow BeCl_2 + H_2$

   42.1%

BALANCE CHEMICAL EQUATIONS

1. Balance the following chemical equations.
   a. $CH_4 + \mathbf{2}O_2 \rightarrow CO_2 + \mathbf{2}H_2O$
   b. $La_2(CO_3)_3 + \mathbf{3}H_2SO_4 \rightarrow La_2(SO_4)_3 + \mathbf{3}H_2O + \mathbf{3}CO_2$
   c. $\mathbf{2}Al(NO_3)_3 + \mathbf{3}Na_2CO_3 \rightarrow Al_2(CO_3)_3 + \mathbf{6}NaNO_3$

2. According to the following unbalanced reactions:
   a. Balance the equations
   b. Determine the amount of reactants and products (in moles)
   c. Explain the phase labels

   i. $KClO_4(s) \rightarrow KCl(s) + 2O_2(g)$

   ii. $BaCl_2(aq) + (NH_4)2CO_3(aq) \rightarrow BaCO_3(s) + 2NH_4Cl(aq)$

   iii. aluminum metal reacts with phosphoric acid to produce hydrogen gas and solid aluminum phosphate.
   $2Al(s) + 2H_3PO_4(aq) \rightarrow 3H_2(g) + 2AlPO_4(aq)$

3. What is the coefficient for H_2 when the equation $Ba + H_3AsO_4 \rightarrow H_2 + Ba_3(AsO_4)_2$ Is properly balanced?
   a. 1
   b. 3
   c. 5
   d. 2
   e. 4

4. Calcium combines with bromine to make calcium bromide. Write the balanced chemical equation for the reaction. What is the coefficient for bromine?
   a. 1
   b. 2
   c. 3
   d. 4
   e. 5

5. According to the following reaction:
   $2Mg(s) + O_2(g) \rightarrow 2MgO(s)$
   What is the phase of the product?
   a. solid
   b. gas
   c. solid and gas
   d. liquid
   e. liquid and gas

6. Barium peroxide, BaO_2, breaks down into barium oxide and oxygen. Write the balanced chemical equation for this reaction. What is the coefficient for barium oxide?
   a. 3
   b. 1
   c. 5
   d. 6
   e. 2

7. Lithium combines with oxygen to form lithium oxide. Write the balanced chemical equation for this reaction. What is the coefficient for lithium?
   a. 4
   b. 3
   c. 5
   d. 1
   e. none of the above

8. Interpret the following sentence:
   "Sodium bicarbonate reacts with acetic acid to produce sodium acetate, carbon dioxide and water."
   a. $Na_2CO_3 + H_2C_2H_3O_2 \rightarrow H_2 + CO_3 + Na_2C_2H_3O_2$
   b. $NaCO_3 + HC_2H_3O_2 \rightarrow NaHC_2H_3O_2 + CO_2 + H_2O$
   c. $NaHCO_3 + H_2O + CO_2 \rightarrow NaC_2H_3O_2 + 2O_2$
   d. $NaHCO_3 + HC_2H_3O_2 \rightarrow NaC_2H_3O_2 + H_2O + CO_2$

CLASSIFY CHEMICAL EQUATIONS

1. Classify the following chemical reactions as:

combination	decomposition	single replacement
double replacement	neutralization	combustion

   a. $Fe(s) + 2HCl \rightarrow H_2(g) + FeCl_2$ **single replacement**
   b. $4NH_3 + 5O_2(g) \rightarrow 4NO + 6H_2O$ **combustion**
   c. $C_2H_2 + HCl \rightarrow C_2H_3Cl$ **combination**
   d. $NaBr(aq) + AgNO_3(aq) \rightarrow AgBr(s) + NaNO_3(aq)$ **double replacement**
   e. $C_6H_{12}O_6 \rightarrow 2C_2H_5OH + 2CO_2$ **decomposition**
   f. $NaOH + CH_3COOH \rightarrow NaCH_3COO + H_2O$ **neutralization**

2. The decomposition by heating of solid potassium chlorate yields solid potassium chloride and oxygen gas as products. Write a balanced equation for this reaction.
   a. $KClO_4 \rightarrow KCl(s) + 2O_2(g)$
   b. $2KClO_3(s) \rightarrow 2KCl(s) + 3O_2(g)$
   c. $KClO_3(s) \rightarrow KCl(s) + 3O(g)$
   d. $2KClO_3(s) \rightarrow 2KClO_2(s) + O(g)$
   e. $KClO_2(s) \rightarrow KCl(s) + O_2(g)$

3. $2Al + 3Sn(NO_3)_2 \rightarrow 2Al(NO_3)_3 + 3Sn$ This equation is an example of which type of reaction?
   a. single replacement
   b. double replacement
   c. combination
   d. decomposition

4. $6K_2O + P_4O_{10} \rightarrow 4K_3PO_4$ This equation is an example of which type of reaction?
   a. single replacement
   b. double replacement
   c. combination
   d. decomposition

5. In class a double displacement reaction was done as a demonstration. A solution of potassium iodide and a solution of lead (II) nitrate were combined. A yellow precipitate formed as a product. What was the precipitate?
   a. KI
   b. $Pb(NO_3)_2$
   c. KNO_3
   d. PbI_2
   e. KPb

6. $2H_{2(g)} + CO_{(g)} \rightarrow CH_3OH_{(l)}$ This equation is an example of which type of reaction?
   a. single replacement
   b. double replacement
   c. combination
   d. decomposition

7. Based upon the type of reaction, determine the products. Write the balanced chemical equation.
   a. Combination: Calcium and bromine combine.
   $Ca(s) + Br_2(l) \rightarrow CaBr_2(s)$

   b. Decomposition: Water breaks apart by electrolysis.
   $2H_2O(g) \rightarrow 2H_2(g) + O_2(g)$

   c. Combustion: Natural gas (CH_4) is burned in furnaces.
   $CH_4(g) + 2O_2(g) \rightarrow CO_2(g) + 2H_2O(l)$

   d. Single Replacement: Chlorine gas is bubbled through an aqueous solution of potassium iodide.
   $Cl_2(g) + 2KI(aq) \rightarrow 2KCl(aq) + I_2(g)$

   e. Double Replacement: Calcium nitrate and potassium fluoride combine to form a precipitate.
   $Ca(NO_3)_2(aq) + 2KF(aq) \rightarrow CaF_2(s) + 2KNO_3(aq)$

   f. Double Replacement: Neutralization: Sodium hydroxide is added to phosphoric acid.
   $3NaOH(aq) + H_3PO_4(aq) \rightarrow Na_3PO_4(aq) + 3H_2O(l)$

STOCHIOMETRY MOLE TO MOLE CONVERSIONS

1. Balance the following equation: _3_ CCl_4 + _2_ SbF_3 → _3_ CCl_2F_2 + _2_ $SbCl_3$

2. According to the following chemical reaction:
   _2_ $Al(s)$ + _3_ $H_2SO_4(aq)$ → ___ $Al_2(SO_4)_3(aq)$ + _3_ $H_2(g)$
   a. Balance the equation.
   b. Indicate the amount of moles reacting and moles being produced of each reactant and product.
   2 moles Al react with 3 moles H_2SO_4 to produce 1 mole $Al_2(SO_4)_3$ and 3 moles H_2

3. According to the following chemical reaction:
$3CaCO_3(s) + 2H_3PO_4(aq) \rightarrow Ca_3(PO_4)_2(aq) + 3H_2O(l) + 3CO_2(g)$
   a. Calculate the moles of calcium phosphate being produced from 45.6 moles of phosphoric acid reacting.
   22.8 mol $Ca_3(PO_4)_2$

   b. Calculate the moles of calcium carbonate reacting if 0.344 moles of water are produced.
   0.344 mol $CaCO_3$

4. How many moles of carbon dioxide are in the following chemical reaction if 2 moles of C_4H_{10} are reacting?
$2C_4H_{10}(g) + 13O_2(g) \rightarrow 8CO_2(g) + 10H_2O(g)$
8 mol CO_2

5. How many moles of water are being produced from 4 moles C_4H_{10} and 26 moles of O_2 reacting?
____ $C_4H_{10}(g)$ + ____ $O_2(g) \rightarrow$ ____ $CO_2(g)$ + ____ $H_2O(g)$
20. mol H_2O

6. How many moles of oxygen are in the following chemical reaction?
____ $Na_2CO_3(aq)$ + ____ $HNO_3(g) \rightarrow$ ____ $H_2O(l)$ + ____ $CO_2(g)$ + ____ $NaNO_3(aq)$
9 mol O_2

7. Calculate the number of moles of Fe_3O_4 produced from 0.75 moles of Fe by the following reaction.
____ Fe + ____ H_2O ____ Fe_3O_4 + ____ H_2
0.25 mol Fe_3O_4

8. Calculate the number of moles of NO produced from 0.25 moles of O_2 by the following reaction.
____ NH_3 + ____ O_2 ____ NO + ____ H_2O
0.20 mol NO

STOCHIOMETRY MOLE TO GRAM CONVERSIONS

1. According to the following chemical reaction:
$Al_2O_3(s) + 6HNO_3(aq) \rightarrow 2Al(NO_3)_3(aq) + 3H_2O(l)$
Calculate the mass of nitric acid if 5.60 moles of water are produced.
705.6 g HNO_3

2. What mass of CaO could be obtained from the thermal decomposition of 2.00 moles of $CaCO_3$?
____ $CaCO_3 \rightarrow$ ____ CaO + ____ CO_2
112 g CaO

3. How many grams of aluminum bromide are formed by the reaction of 1.50 moles of HBr according to the following equation?

_____ Al + _____ HBr → _____ $AlBr_3$ + _____ H_2

733 g $AlBr_3$

4. How many grams of H_2 are produced by the reaction of 0.256 mol of H_3PO_4 according to the following equation?

_____ Cr + _____ H_3PO_4 → _____ $CrPO_4$ + _____ H_2

0.768 g H_2

STOCHIOMETRY GRAM TO MOLE CONVERSIONS

1. According to the following chemical reaction:

   $Al_2O_3(s) + 6HNO_3(aq) \rightarrow 2Al(NO_3)_3(aq) + 3H_2O(l)$

   Calculate the moles of aluminum oxide if 43.3 grams of aluminum nitrate are produced.

   0.102 mol Al_2O_3

STOCHIOMETRY GRAM TO GRAM CONVERSIONS

1. According to the following chemical reaction:

   $Al_2O_3(s) + 6HNO_3(aq) \rightarrow 2Al(NO_3)_3(aq) + 3H_2O(l)$

   a. Calculate the mass of water produced if 0.430 grams of aluminum nitrate are produced.

   0.0545 g H_2O

   b. Calculate the mass of aluminum oxide reacting if 88.8 grams of water are produced.

   168 g Al_2O_3

2. Freshly exposed aluminum surfaces react with oxygen to form a tough oxide coating that protects the metal from further corrosion. How many grams of O_2 are required to react with 8.09 g of Al?

   _____ Al + _____ O_2 → _____ Al_2O_3

   7.19 g O_2

3. Barium chloride was used to precipitate silver chloride from a solution of silver nitrate. What mass of barium chloride had to react if 0.635 grams of silver chloride formed?

   _____ $BaCl_2(aq)$ + _____ $AgNO_3(aq)$ → _____ $AgCl(s)$ + _____ $Ba(NO_3)_2(aq)$

   0.461 g $BaCl_2$

STOCHIOMETRY COMBINED CONVERSIONS

1. According to the following reaction:

   _____ WO_3 + _____ H_2 → _____ W + _____ H_2O

   How many moles of tungsten are produced by the reaction of 0.00761 mole of WO_3 with hydrogen?

   0.00761 mol W

2. According to the following reaction:

   ____ $Mg(s)$ + ____ $O_2(g)$ → ____ $MgO(s)$

   How many moles of MgO are produced of 4.60 moles of oxygen react?

   9.20 mol MgO

3. According to the following reaction:

   ____ N_2 + ____ O_2 → ____ NO

   How many moles of NO produced by the reaction of 26.8 grams of nitrogen?

   1.91 mol NO

4. According to the following reaction:

   ____ $CH_4(g)$ + ____ $O_2(g)$ → ____ $CO_2(g)$ + ____ $H_2O(g)$

   How many moles of NO produced by the reaction of 26.8 grams of nitrogen?

   1.00 mol CH_4

5. According to the following reaction:

   ____ $Al(s)$ + ____ $O_2(g)$ → ____ $Al_2O_3(s)$

   How many grams of oxygen react with 108 grams aluminum?

   384 g O_2

6. According to the following reaction:

   ____ CCl_4 + ____ SbF_3 → ____ CCl_2F_2 + ____ $SbCl_3$

   If 4.36 grams of Freon-12 (CCl_2F_2) is produced in the reaction, how many grams of $SbCl_3$ are also produced?

   5.48 g $SbCl_3$

STOICHIOMETRY LIMITING REACTANT

1. If 25.0 grams of Al_2O_3 and 75.0 grams of carbon react according to the following chemical reaction:

   $Al_2O_3(s) + 3C(s) \rightarrow 2Al(s) + 3CO(s)$

   a. Which is the limiting reactant and which is the excess reactant?

      Al_2O_3

   b. What is the maximum number of grams of Al that can be produced?

      13.2 g Al

   c. What is the mass of excess reactant remaining after the reaction?

      66.2 g C

2. According to the following reaction:

   ____ Sb + ____ Cl_2 → ____ $SbCl_3$

   If Sb is completely consumed in the reaction, it is the (limiting or excess) reactant?

   Limiting

3. What mass of $PbSO_4$ is produced when 1.94 g $Pb(NO_3)_2$ reacts with 0.83 g $Al_2(SO_4)_3$?

____ $Pb(NO_3)_2$ + ____ $Al_2(SO_4)_3$ → ____ $PbSO_4$ + ____ $Al(NO_3)_2$

1.78 g $PbSO_4$

4. Cu_2HgI_4 is prepared according to the equation:

$2CuI + HgI_2 \rightarrow Cu_2HgI_2$

When 2.00 grams of each reactant are used, which one is the limiting reactant?

HgI

5. If the reaction $N_2 + 3H_2 \rightarrow 2NH_3$ is carried out using 1.40 grams of N_2 and 0.400 grams of H_2, what mass of excess reactant will remain?

0.100 g H_2

PERCENTAGE YIELD CALCULATIONS

1. According to the following chemical reaction:

____ Na_2CO_3 + ____ HCl → ____ NaCl + ____ H_2O + ____ CO_2

5.00 grams of Na_2CO_3 was treated with excess HCl, 1.50 grams of CO_2 was obtained (actual yield).

a. Calculate the theoretical amount of grams of CO_2 produced.

2.08 g CO_2

b. Calculate the percentage yield of CO_2.

72.3%

2. Reaction of 1.00 mole CH_4 with excess Cl_2 yields 96.8 g CCl_4 (actual yield). What is the percentage yield of the reaction?

____ CH_4 + ____ Cl_2 → ____ CCl_4 + ____ HCl

62.9% CCl_4

3. Toluene is oxidized by air under carefully controlled conditions to benzoic acid which is used to prepare the food preservative sodium benzoate. What is the yield of a reaction in which 1.00 g of toluene is converted to 1.21 g of benzoic acid (actual yield)?

$2C_6H_5CH_3 + 3O_2 \rightarrow 2C_6H_5CO_2H + 2H_2O$

91.2%

4. The reaction of 6.8 g of H_2S with excess SO_2 according to the following reaction yields 8.2 g of S (actual yield). What is the percentage yield?

____ H_2S + ____ SO_2 → ____ S + ____ H_2O

85.4% S

5. Which is the best percentage yield?
   a. 0%
   b. 10%
   c. 50%
   d. 75%
   e. 100%

6. In a general chemistry laboratory experiment, a student produces 2.73 grams of a compound. She calculates the theoretical yield as 3.40 grams. What is the percentage yield?

   80.3%

PERCENTAGE YIELD WITH LIMITING REACTANT CALCULATIONS

1. The mass of Li_2O formed when 2.00 g of lithium reacts with 2.00 g of oxygen is 3.02 g (actual yield). What is the percentage yield?

   ____ Li + ____ O_2 → ____ Li_2O

   81.1% Li_2O

2. The mass of S_2Cl_2 formed when 6.00 g of sulfur reacts with 6.00 g of chlorine is 9.5 g (actual yield). What is the percentage yield?

   ____ S_8 + ____ Cl_2 → ____ S_2Cl_2

3. The mass of iron produced by the reaction of 7.00 g of Fe_2O_3 and 3.00 g of CO is 3.55 g (actual yield). What is the percentage yield?

   $Fe_2O_3 + 3CO \rightarrow 2Fe + 3CO_2$

   89.1%

4. If 2.75 grams of NaI are produced from a mixture initially containing 5.00 grams I_2 and 1.00 grams of NaOH, what is the percentage yield?

   $3I_2 + 6NaOH \rightarrow 5NaI + NaIO_3 + 3H_2O$

   88.2%

UNIT 8: PRACTICE PROBLEMS

Convert Units for Gas Laws

1. Convert 744 torr to atmospheres and to kPa.

   0.979 atm / 99.2 kPa

2. How many torr is equivalent to 1.25 atm?

   950. torr

3. What is 20.5°F to Kelvin?

   –6.4°C / 266.8 K

4. Convert 357 mL to liters.

   0.357 L

5. How many liters are in 1.125 grams of carbon dioxide gas at STP?

   0.5727 L

Combined Gas Laws

6. If an amount of gas at 0.978 atm and 0.456 L is compressed at a constant temperature to a volume of 0.250 L, what is the pressure?

   1.78 atm

7. What volume would a hydrogen balloon be at 725 torr, if the temperature is constant and it had a volume of 1.527 L at 759 torr?

   1.60 L

8. An aerosol can has a pressure of 1.30 atm at 25°C. What is the pressure at 200°C? At constant volume?

   2.06 atm

9. A diver set a tanks pressure to 770.25 torr at 20°C, in the water the pressure changes to 745.6 torr what is the temperature of the tank in the water? At constant volume?

   284 K or 11°C

10. Given a 526 mL sample of a gas at 22.7°C and 748.9 torr, what is the volume if the temperature and pressure are at STP?

    478 mL

11. A gas has a volume of 0.00400 mL at 25.5 atm and 25°C. What is the pressure if the volume of the gas changed to 8.00 mL and the temperature changed to 55°C?

    0.0140 atm

12. A gas pressure at 781 torr at 20°C in a 456 mL tank has a pressure change to 745.6 torr. What is the temperature of the tank in the water at constant volume?

    280. K or 7°C

13. A hot air balloon has a volume of 1600 L at 180°C. What is the volume if the gas is heated to 220°C?

    1741 L = 1700 L

14. At 293.6 K a gas has a volume of 356.3 mL. What is the temperature if the volume changes to 405.8 mL?

    334 K

GAS LAWS: IDEAL GAS

1. If I have 4 moles of a gas at a pressure of 5.6 atm and a volume of 12 liters, what is the temperature?
   205 K, –68°C

2. If I have 21 moles of gas held at a pressure of 3800 torr and a temperature of 627°C, what is the volume of the gas?
   310 L

3. If I have 54.3 grams of nitrogen held at a pressure of 5 atm and in a container with a volume of 50 liters, what is the temperature (in Celcius) of the gas?
   1578 K, 1305°C

4. Calculate the molecular mass of 4.5 grams of a gas at a volume of 345 mL and a pressure of 45.8 psi at 25°C.
   102 g/mol

5. A gas exerts a pressure of 0.892 atm in a 523 mL container at 15°C. The density of the gas is 1.22 g/L. What is the molecular mass of the gas?
   32.3 amu

6. Determine the density of helium at STP.
   0.178 g/L

ASSOCIATION WITH GASES: PRESSURE, VOLUME, TEMPERATURE, AND MOLES

1. Perform the following conversions:
   a. 453 kPa = **4.47** atm
   b. 281.3 cm^3 = **0.2813** L
   c. 61°C = **142** °F = **334** K
   d. 92.1 g Cl_2 = **1.30** molCl_2

2. Calculate the following volume at STP for each:
   a. 0.45 mol N_2 = **10** L N_2
   b. 2.3×10^4 mol Ar = **5.2×10^8** mL Ar
   c. 92.1 g Cl_2 = **29.1** L Cl_2

3. The pressure of nitrogen in a cylinder is 2.00×10^3 psi. What is the pressure in atmospheres?
   136 atm

4. The vapor pressure of mercury is 0.0012 torr at 20°C. What is this pressure in atmospheres?
   1.58×10^{-6} atm

COMBINED GAS LAW

1. Use the gas laws to calculate the following:
   a. 1.2 L of carbon disulfide is heated from 15°C to 80°C at constant pressure. What is the new volume?
   1.5 L

   b. 155 mL of freon is taken from 810 torr to standard pressure. What is the new volume at constant temperature?
   165 mL

   c. 4.12 L of oxygen is at 0.977 atmospheres and 15.2°C. What is the volume at 1.04 atmospheres and 42.0°C?
   4.23 L

2. Consider a cylinder fitted with a movable piston. The initial pressure inside the cylinder is P_i and the initial volume is V_i. What is the new pressure in the system when the piston decreases the volume of the cylinder by half?
   a. $2\ V_iP_i$
   b. $(1/4)\ P_i$
   c. P_i^2
   d. $2\ P_i$
   e. $(1/2)\ P_i$

3. The pressure of a combustible mixture in a cylinder of a motorcycle engine is 0.980 atm when the volume is 246 mL. The piston decreases the volume to 24.1 mL. What is the pressure (atm) at that point assuming no change in temperature occurs?
   10.0 atm

4. What would be the volume (mL) of a sample of ethane (C_2H_6) at 467 K and 1.2 atm if it occupied 405 mL at 298 K and 1.2 atm?
   258 mL

5. A balloon is filled with air and has a volume of 3.25 L at 30°C. The balloon is placed in a freezer at −10°C. What is the volume of the balloon at this temperature?
   2.82 L

6. A sample of a gas at −91°C and 1 atm occupies 2.0 L. What volume, L, will the gas occupy at 0°C at the same pressure?
   3.0 L

7. A spray can is used until only the propellant gas remains at a pressure of 1.1 atm at 23°C. If the can were thrown into a fire at 475°C, what would be the pressure (atm) in the hot can?
   2.78 atm

8. Gas evolved in the fermentation of sugar in wine making occupies a volume of 0.75 L at 20°C at 720 mm Hg. What volume (L) would the gas occupy at 39°C and 1.00 atm?

   0.76 L

9. A gas occupies 1.0 L at 27°C and 0.50 atm. At what temperature (°C) will the gas occupy at 0.50 L at 1.0 atm?

   27°C

IDEAL GAS LAW

1. Use the ideal gas law to calculate the following:
   a. What pressure will be exerted by 1.42 moles of butane, C_4H_{10}, in a 10.0 L cylinder at 75°C?

      4.06 atm

   b. A sample of oxygen is collected in a laboratory experiment. Find its mass if the volume is 618 mL at 790 torr and 15°C.

      0.870 g O_2

2. A 25 L cylinder contains 128 g of nitrogen gas at 10°C. How many grams of nitrogen must be added to increase the pressure to 5.00 atm assuming ideal gas behavior?

   151 g

3. A balloon is filled with 48.3 g of helium at 31°C and 2.12 atm. What is the volume of the balloon in liters?

   142 L

4. A 40.0 gram sample of helium is introduced into a 2.24 L cylinder which is heated until the pressure is 200 atm. What is the temperature (in °C) of the gas?

   273°C

5. An incandescent light bulb with a volume of 125 cm^3 contains 2.5×10^{-3} moles of argon. What is the pressure of argon (atm) at 25°C?

   0.489 atm

6. The pressure in a 2.0 L container is 1.5×10^{-4} torr at 1115 K. How many moles are in the container?

   4.3×10^{-9} mol

7. The Goodyear blimp has 5.12×10^6 liters of helium at 25°C and 1.00 atm. What mass, g, of helium is in the blimp?

   8.37×10^5 g

8. A 725 gram sample of neon is introduced into a 4.5 L cylinder which is then heated until the pressure is 225 atm. What is the temperature (°C) of the gas?
   70.°C

9. What volume is occupied by 0.0100 mole of carbon monoxide at 9°C and 0.973 atm?
   0.238 L

10. What is the temperature of 0.0250 mole of hydrogen gas if it occupies 665 mL at 715 torr?
    305 K

MOLECULAR MASS AND DENSITY

1. Use the molecular mass and density formulas to calculate the following:
   a. 4.22 grams of a gas has a volume of 2.32 liters and a pressure of 0.211 atm. If the temperature is 25.0°C, calculate the molar mass.
      211 g/mol

   b. What is the density in g/L of BF_3 at STP?
      3.03 g/L

   c. A gaseous compound was found to have a density of 5.60 g/L at 23.0°C ant 750 torr. Calculate the molar mass.
      138 g/mol

2. What is the density (g/L) of nitrogen at STP?
   1.25 g/L

3. What is the density in g/L of BrF_3 at 425 torr and 77°C?
   2.66 g/L

4. Calculate the density of neon at 32°C and 0.676 atm.
   0.545 g/L

5. Calculate the density of nitrogen dioxide at STP.
   2.05 g/L

GAS STOICHIOMETRY

1. Using stoichiometry, calculate the following:
   a. What volume of oxygen gas at STP is produced by the decomposition of 100.0 grams of sodium nitrate?

      $2NaNO_4 \rightarrow 2NaNO_3 + O_2$

      13.18 L O_2

   b. How many liters of CO_2 are produced at 25.0°C at 760. torr by burning 24.0 grams of CH_4 in oxygen?

      $CH_4 + 2O_2 \rightarrow CO_2 + 2H_2O$

      36.7 L CO_2

2. What volume of hydrogen (in L) at STP would be required to react with 0.100 mole of nitrogen to form ammonia?

   $N_2(g) + 3H_2(g) \rightarrow 2NH_3(g)$

   6.72 L H_2

UNIT 9: PRACTICE PROBLEMS

MOLARITY

1. Sea water contains roughly 28.0 g of NaCl per 1.00 liter. What is the molarity of sodium chloride in sea water?
   0.479 M NaCl

2. What is the molarity of 245.0 g of H_2SO_4 dissolved in 1.000 L of solution?
   2.498 M H_2SO_4

3. What is the molarity of 5.30 g of Na_2CO_3 dissolved in 400.0 mL solution?
   0.125 M Na_2CO_3

4. How many moles of Na_2CO_3 are in 10.0 mL of a 2.0 M solution?
   0.020 mol Na_2CO_3

5. What mass (in grams) of H_2SO_4 would be needed to make 750.0 mL of 2.00 M solution?
   147 g H_2SO_4

6. What volume (in mL) of 18.0 M H_2SO_4 is needed to contain 2.45 g H_2SO_4?
   1.39 mL

7. What is the molarity of a solution made by dissolving 20.0 g of H_3PO_4 in 50.0 mL of solution?
   4.08 M

DILLUTIONS

8. A stock solution of 1.00 M NaCl is available. How many milliliters are needed to make 100.0 mL of 0.750 M

   75.0 mL

9. What volume of 0.250 M KCl is needed to make 100.0 mL of 0.100 M solution?

   40.0 mL

10. Concentrated H_2SO_4 is 18.0 M. What volume is needed to make 2.00 L of 1.00 M solution?

    0.111 L

11. Concentrated HCl is 12.0 M. What volume is needed to make 2.00 L of 1.00 M solution?

    0.167 L

12. A 0.500 M solution is to be diluted to 500.0 mL of a 0.150 M solution. How many mL of the 0.500 M solution are required?

    150. mL

SOLUTIONS

1. Give an example of a homogeneous solution. Indicate the solute and solvent. Does this solution contain ions? If yes, indicate what the ions are. If no, how does the solute dissolve?

   Any salts dissolved in water. Such as NaI. The solute ins the NaI, the solvent is the water. The ions are Na^+ and I^-.

2. Illustrate how each of the following compounds dissolves in water.

   a. $BaCl_2$

   IONIC COMPOUND DISSOLVED IN WATER. $BaCl_2$ break apart into ions, in the following proportion Ba_2^+ and $2Cl^-$. These become HYDRATED IONS because they are surrounded by water molecules.

   b. Ethanol, C_2H_5OH

   COVALENT COMPOUND DISSOLVED IN WATER. C_2H_5OH is "surrounded" by water molecules, which keep C_2H_5OH in solution. These become SOLVATED COMPOUNDS.

   c. $Mg(ClO_4)$

   IONIC COMPOUND DISSOLVED IN WATER. $Mg(ClO_4)_2$ break apart into ions, in the following proportion Mg_2^+ and $2ClO_4^-$. These become HYDRATED IONS because they are surrounded by water molecules.

   d. Perchloric acid, $HClO_4$ (a strong acid)

   COVALENT COMPOUND DISSOLVED IN WATER. $HClO_4$ is "pulled apart" into separate ions by the strong charges on the water molecules. The shared bonds actually get broken. These become IONIZED (UNDERGO IONIZATION).

3. Explain the differences between saturated and super saturated solutions.

   A good analogy is to compare a solution to a movie theater. If the seats represent the solvent and the people represent the solute... In a saturated situation, all the seats are filled in the theater, with one person per seat. In a super saturated solution, all the seats are filled in the theater with more than one person per seat.

CONCENTRATION CALCULATIONS

4. Calculate the percent by mass of solute if 14.15 grams of NaI is mixed with 75.55 grams of water.

   15.77% NaI

5. A solution contains 15.0 grams of NH_4Cl in water and 8.50% NH_4Cl. What is the mass of water present?

   161 g H_2O

6. What is the molarity of a solution of 345 grams of Epsom salts ($MgSO_4 \bullet 7H_2O$) in 7.50 L of solution?

   0.187 M $MgSO_4 \bullet 7H_2O$

7. What is the volume in liters of a 0.440 *M* solution if it was made by dilution of 250 mL of a 1.25 *M* solution?

   0.710 L

8. Calculate the grams of alcohol, C_2H_5OH, in 440 grams of a 23.0% solution.

   101 g

9. 75.0 mL of water is added to 12.9 mL of 0.250 M $KMnO_4$. What is the concentration of the diluted solution?

   0.0367 M

10. If 2.54 g of sodium benzoate is dissolved in 75.00 g H_2O. What is the percentage concentration of sodium benzoate?

    3.28%

11. If 45.5 g of $BaCl_2$ is dissolved in water to produce 2.74 L of solution, what is the molarity of the solution?

    0.0797 M $BaCl_2$

12. Nitric acid is commercially available at a concentration of 15.9 M. What volume of this solution must be diluted to a final volume of 1.00 L to prepare a 4.00 M solution?

    0.25 L

SOLUTION STOICHIOMETRY

13. Consider the reaction:
    a. 3 $Ca(ClO_3)_2(aq)$ + 2 $Na_3PO_4(aq) \rightarrow Ca_3(PO_4)_2(aq)$ + 6 $NaClO_3(aq)$
    b. What volume of a 2.22 *M* solution of Na_3PO_4 is needed to react with 580 mL of a 3.75 *M* solution of $Ca(ClO_3)_2$?
    0.653 L

14. The citric acid in a lemon juice sample was neutralized by titration with NaOH solution. If 5.00 mL of lemon juice required 47.8 mL of 0.121 M NaOH for neutralization, what was the molarity of the citric acid in the lemon juice?

    0.386 M

    The reaction is ______ NaOH + ______ $H_3C_6H_5O_7$ $\rightarrow$ ______ H_2O + ______ $Na_3C_6H_5O_7$

IDENTIFYING ACIDS AND BASES

15. Identify each of the following as an acid or base when dissolved in water.

a. HClO	**ACID**	name	**hypochlorous acid**
b. $Mn(OH)_2$	**BASE**	name	**manganese (II) hydroxide**
c. H_2Se	**ACID**	name	**selenic acid**
d. H_2SO_4	**ACID**	name	**sulfuric acid**
e. CuOH	**BASE**	name	**copper (I) hydroxide**

16. Balance each of the following neutralization reactions.
    a. __**2**__ $HNO_3 + Ca(OH)_2 \rightarrow Ca(NO_3)_2$ + __**2**__ H_2O
    b. $HC_2H_3O_2 + NaOH \rightarrow NaC_2H_3O_2 + H_2O$
    c. $H_2SO_4 + Mg(OH)_2 \rightarrow MgSO_4$ + __**2**__ H_2O

17. Identify the acid, base, conjugate acid, conjugate base for each of the following reactions.

a. $HClO_3$	+	H_2O	$\rightarrow$	ClO_3^-	+	H_3O^+
acid	+	**base**	$\rightarrow$	**c. base**	+	**c. acid**
b. HSO_4^-	+	NH_3	$\rightarrow$	NH_4^+	+	SO_4^{2-}
acid	+	**base**	$\rightarrow$	**c. acid**	+	**c. base**
c. HCO_3^-	+	H_2O	$\rightarrow$	H_2CO_3	+	OH^-
base	+	**acid**	$\rightarrow$	**c. acid**	+	**c. base**

pH AND pOH

18. What is the $[H^+]$ in the following solutions?
    a. $[OH^-] = 1.0 \times 10^{-12}$ **$[H^+] = 1.0 \times 10^{-2}$**
    b. $[OH^-] = 5.9 \times 10^{-4}$ **$[H^+] = 1.7 \times 10^{-11}$**
    c. pH = 9.0 **$[H^+] = 1.0 \times 10^{-9}$**
    d. pH = 7.85 **$[H^+] = 1.4 \times 10^{-8}$**
    e. pOH = 4.18 **$[H^+] = 1.5 \times 10^{-10}$**

19. What is the pH of the following solutions?
   a. pOH = 4.51 **pH = 9.49**
   b. $[OH^-] = 1.0 \times 10^{-7}$ **pH = 8.00**
   c. $[H^+] = 3.6 \times 10^{-4}$ **pH = 3.44**

20. Which of the following is a strong acid?
   a. H_2CO_3
   b. $HC_2H_3O_3$
   c. HF
   d. HCl
   e. HClO

21. Which of the following is a characteristic property of a traditional acid?
   a. Turns litmus indicator blue
   b. Feels slippery
   c. Tastes bitter
   d. Sour taste
   e. Contain OH^- ions

22. Which one is a common indicator used in acid base titrations?
   a. sodium chloride
   b. propylene
   c. phenolphthalein
   d. bronsted-lowry
   e. arrhenius

23. According to Arrhenius theory, what is an acid?
   a. a substance which contains a high concentration of hydrogen ions in solutions with water
   b. a substance which will lower the hydrogen ion concentration when placed in water a substance that has an h in its formula
   c. an electron pair donor
   d. an electron pair acceptor

24. A Brønsted-Lowry acid is defined as a(n) ______.
   a. proton acceptor
   b. electron donor
   c. proton donor
   d. electron pair acceptor
   e. electron pair donor

25. What is a Brønsted-Lowry base?
   a. an electron pair donor
   b. a hydroxide ion donor
   c. a metal ion
   d. a proton donor
   e. a proton acceptor

26. What kind of substance is water in the following reaction?
   $H_2O(s) + HCN(aq) \rightarrow H_3O^+(aq) + CN^-(aq)$
   a. A Brønsted-Lowry base
   b. A Brønsted-Lowry acid